Polymer Science and Technology - Formulae Masterbook

A Comprehensive, Categorized Collection of Definitions and Equations used in Polymer Science and Surface Coating Technology

Dr. Kunal K Wadgaonkar

Dr. Arjit A Gadgeel

notionpress.com

INDIA • SINGAPORE • MALAYSIA

Made with ♥ on the Notion Press Platform
www.notionpress.com

Preface

Polymer Technology is a multidisciplinary field that encompasses the synthesis, processing, characterization, and application of polymeric materials. It integrates principles of chemistry, physics, and engineering to design and manufacture materials with tailored properties for a wide range of industries, including automotive, medical devices, packaging, coatings, construction, electronics, and advanced manufacturing. Continuous innovation in polymer science enables the development of materials with improved performance, durability, sustainability and functionality.

Understanding polymer systems requires familiarity with numerous fundamental definitions, equations, and relationships that govern polymerization kinetics, molecular structure, thermodynamics, rheology, surface interactions, and coating performance. These relations form the foundation for scientific analysis, process optimization, product development, and troubleshooting in industrial application.

The presented collection is an effort to provide easy access to the information regarding the definitions and equations used in the field of Polymer Science and Coating Technology. Designed for students, researchers, formulators, manufacturers, and technologists the handbook serves as an easily accessible source of information.

All the sources from where the data was compiled are mentioned at the end in the bibliography. This artwork represents our continuous efforts to ease the process of education and innovation. Please reach us at contact@polymerdna.com if some important data has been missed by us as continuous development is our primary goal.

About PolymerDNA

PolymerDNA is a platform dedicated to accelerating innovation in the world of polymers through Development, Networking and Analysis (DNA).

PolymerDNA steers its way with a vision to bridge the gap between scientific research and industrial application by offering unique blend of technical insights and articles tailored for the polymer and surface coating industry.

At PolymerDNA, we believe that the future of material science lies in collaboration and continuous learning. Our platform serves as a hub for professionals to stay ahead in the rapidly evolving field through its content, insightful commentary, and an active network of professionals.

We at PolymerDNA strive hard to provide you concise, clear-cut answers to real problems faced by researchers, formulators, manufacturers and innovators via blogs covering innumerable topics. Visit our website polymerdna.com for more information.

Table of Contents

01

Polymer Chemistry and Stoichiometry

Polymer Chemistry & Stoichiometry

Molecular Weight:

The number-average molecular weight (Mn) is the total mass of material divided by the total number of molecules, and is therefore a numerical average. Properties such as vapour-pressure lowering, freezing-point depression and osmotic pressure depend on the number of molecules, not on their size.

The weight-average molecular weight (Mw) is based on the assumption that the individual molecule contributes to the total weight in proportion to its mass, rather than its number. Properties like diffusion, sedimentation, light scattering depend on both molecular size and mass.

Apart from Mn and Mw, other molecular-weight expressions include the z-average (Mz) and viscosity-average (Mv), derived from sedimentation and flow behaviour in solution.

Number-Average Molecular Weight:

$$M_n = \frac{\sum \mathrm{n}_i \mathrm{M}_i}{\sum \mathrm{n}_i}$$

Weight-Average Molecular Weight:

$$M_w = \frac{\sum \mathrm{n}_i {\mathrm{M}_i}^2}{\sum \mathrm{n}_i \mathrm{M}_i}$$

Z-Average Molecular Weight:

$$M_z = \frac{\sum \mathrm{n}_i M_i^3}{\sum \mathrm{n}_i M_i^2}$$

Viscosity-Average Molecular Weight:

$$M_v = \left(\frac{\sum \mathrm{n}_i M_i^{1+a}}{\sum \mathrm{n}_i M_i^2}\right)^{1/a}$$

Polydispersity Index (PDI):

Polydispersity describes the broadness of the molecular-weight distribution. It is represented by a curve plotting the number fraction of molecules of a given molecular weight versus that molecular weight.

PDI is an important parameter because it indicates the lowest and highest molecular-weight species, as well as the distribution of intermediate species.

A quick estimate of polydispersity is the Mw/Mn ratio, which indicates distribution breadth.

$$PDI = \frac{\mathrm{M}_w}{\mathrm{M}_n}$$

Following table shows the Mw/Mn for polymers obtained by different polymerisation reactions.

Polymerisation type		**Mw/Mn**
Free radical chain polymerisation	With precise temperature control (solution, suspension and emulsion systems)	1.5-2
	With moderate temperature control (bulk)	2-5
	Without temperature control (with auto-acceleration)	8-10
Cationic/Anionic chain polymerisation	Using homogeneous catalyst system	<1.5
	Using heterogeneous catalyst system	>10
Polycondensation, polyaddition and ring opening polymerisation		2-3
Coordination polymerisation		>10
Polymerisation systems leading to chain-branching		>20

Molecular Weight via Osmometry:

Osmometry determines Mn. It is based on the phenomenon of osmosis, where a solvent diffuses through a semipermeable membrane into a solution due to chemical-potential differences. Applying pressure on the solution side stops the flow.

Thus, osmotic pressure (π) is defined as the pressure required to prevent solvent flow into the solution.

The osmotic pressure of a polymer solution is related to Mn as follows:

$$\frac{\pi}{RTC} = \frac{1}{M_n} + BC$$

$$\left(\frac{\pi}{RTC}\right)_{c\to 0} = \frac{1}{M_n}$$

Where,

π = Osmotic pressure

R = Universal gas constant

T = Absolute temperature

C = Concentration of the solution

Mn = Number average molecular weight

A plot of (π/RTC) versus C gives a straight line whose intercept gives 1/Mn.

Static Light Scattering (Rayleigh):

Static light scattering measures molecular weight from relationships between scattered-light intensity and molecular size. According to Rayleigh theory, molecular weight is proportional to the Rayleigh ratio (scattered intensity / incident intensity).

Key points:

- RALS (Right Angle Light Scattering) detects scattered light at 90°, suitable for isotropic scatterers.
- LALS (Low Angle Light Scattering) detects at ~7°, suitable for all molecules but with lower signal-to-noise.
- Hybrid RALS/LALS improves accuracy.

- MALS (Multi Angle Light Scattering) detects at multiple angles, allowing determination of both MW and radius of gyration for anisotropic scatterers.

The properties of the light scattered by molecules and particles vary depending on the object doing the scattering as defined in the Rayleigh equation.

$$\frac{KC}{R_\theta} = \left(\frac{1}{M_w} + 2A_2C\right)\frac{1}{P_\theta}$$

Where,

C = Concentration

θ = Scattering Angle

R_θ = Rayleigh ratio

M_W = Weight-average molecular weight

A_2 = Second virial coefficient

K and P_θ are more complex terms defined as:

$$K = \frac{4\pi^2}{\lambda_0^4 N_A}\left(n_0\frac{dn}{dc}\right)^2$$

Where,

λ_0 = Laser wavelength in the vacuum

N_A = Avogadro's number

n_0 = Refractive index of the solvent

dn/dc = Refractive index increment of the sample.

$$\frac{1}{P_\theta} = 1 + \frac{16\pi^2 n_0^2 R_g^2}{3\lambda_0^2} Sin^2\left(\frac{\theta}{2}\right)$$

Where,

R_g = Molecule's radius of gyration

Because larger molecules scatter more light, scattered-intensity increases linearly with MW but non-linearly with size.

To overcome angular dependence, scattered intensity is extrapolated to 0° using a Guinier plot (KC/Rθ vs. $\sin^2(\theta/2)$), allowing MW evaluation from the intercept.

Intrinsic Viscosity (Mark-Houwink Equation):

The Mark–Houwink equation describes the empirical relationship between intrinsic viscosity and polymer molecular weight:

$$[\eta] = KM^\alpha$$

Where,

η = Intrinsic viscosity

M = Molecular weight

K, α = Constants for a polymer-solvent system

α (viscosity exponent):

- $\alpha \approx 0$ → compact or spherical structure
- 0.5–0.8 → random coil polymer
- $\alpha > 1$ → rigid, rod-like polymer

K relates to hydrodynamic volume and depends on polymer, solvent, and temperature.

K and α are often taken from literature or derived from a log–log plot of intrinsic viscosity versus molar mass.

K and α are constants for a given polymer-solvent combination at a certain temperature. If they are not known from literature, they have to be determined experimentally by plotting the logarithm of the intrinsic viscosity as a function of the logarithmic molar mass. The intersection with the y-axis of a linear fit of data gives the constant log K, the slope of the curve gives the constant α.

Average Functionality

Average functionality (f_{avg}) for a mixture of monomers or functional oligomers is:

$$favg = \frac{\sum n_i f_i}{\sum n_i}$$

Where,

n_i = Moles of molecule i

f_i = Functionality of molecule i

Tg of Copolymers (Fox Equation)

The Tg of a copolymer can be estimated mathematically using the Fox Equation:

$$\frac{1}{Tg\ (copolymer)} = \frac{W_1}{Tg_1} + \frac{W_2}{Tg_2} + \cdots + \frac{W_n}{Tg_n}$$

Where

W_1, W_2... W_n = Weight fraction

Tg_1, Tg_2... Tg_n = Homopolymer Tgs in Kelvin

For Step Growth Polymerisation

The number-average degree of polymerisation X_n is:

$$Xn = \frac{1 + r}{1 + r - 2rp}$$

$$p = \frac{2\ (N_0 - N)}{N_0 \text{ x } F_{avg}}$$

Where,

p = Extent of reaction

N_o = Initial number of molecules

N = Molecules at time t

r = Stoichiometric ratio (N_a/N_b); r <1.

As per Carother's definition gel point corresponds to an infinite number average molecular weight. Therefore, at gelation $N_o >> N$

Therefore the equation becomes,

$$P_{gel} = \frac{2}{f_{avg}}$$

At completion of reaction p = 1;

$$X_n = \frac{1+r}{1-r}$$

Molecular Weight between Two Crosslinks

The molecular weight of the segment between two crosslinks in a polymer molecule is given by the following equation.

$$M_c = \frac{3\rho RT}{E'}$$

Where,

M_C = Molecular weight between crosslinks

ρ = Polymer density

R = Gas constant

T = Absolute temperature

E' = Storage modulus.

Monomer Conversion / Extent of Reaction:

This simpler expression applies only to monomers with a single functional group (mono-functional species), such as vinyl monomers in chain-growth polymerisation.

$$p = \frac{n_0 - n_t}{n_0}$$

n_0 = Number of moles at the start of reaction

n_t = Number of remaining moles at any time t

02

Fundamentals of Polymer and Surface Coating Technology

02

Fundamentals of Polymer and Surface Coating Technology

Fundamentals of Polymer and Surface Coating Technology

Oil Absorption Value (As Per ASTM D 281-31)

Oil absorption gives the minimum amount of oil required to completely wet a given pigment.

It is expressed as grams of oil/100 gram of sample.

The oil used is alkali refined linseed oil. In the rub out method, the pigment is mixed with the oil till the pigment is wetted, i.e. no dry pigment is noticed. The oil is added by a burette into the pigment until it forms a single mass as a stiff paste of smooth consistency (putty-like paste which doesn't break or separate). The oil absorption figure represents the minimum weight of the oil required to coat each pigment particle and to fill the voids between them. The factors affecting oil absorption value are:

- Pigment type
- Particle size and its distribution
- Shape of the particles
- Pigment surface characteristics
- Wettability by the oil and the pigment itself
- Time required for evaluation
- Method

Oil absorption value is calculated as

$$Oil\ absorption\ value = \frac{Burette\ reading \times sp.gravity\ of\ oil \times 100}{Weight\ of\ pigment}$$

Instead of using a burette, the oil used can be weighted and the amount given directly in grams. The formula then reads

$$Oil\ absorption\ value = \frac{Weight\ of\ oil}{Weight\ of\ pigment}\ x\ 100$$

Hydrophilic-Lipophilic Balance (HLB Value)

HLB value tells us the chemical character of the surfactant. Lower values indicate low water solubility and higher oil solubility. Higher values indicate higher water solubility and lower oil solubility.

It is not possible to exactly demarcate where a de-emulsifier ends and a wetting and dispersing agent starts on the HBL scale or where a dispersing agent ends and an emulsifier starts. The general guideline is

1. De-emulsifier's HLB is in the range of 2 to 4
2. Wetting and dispersing agent's HLB is from 4 to 8
3. Emulsifier's HLB is between 7 to 40

There are several ways in which the HLB value of a given chemical can be determined. In one method, various structural groups are allotted a number that indicates the intensity by which the group affects the water sensitivity or water repellence of a compound. The contribution of both the types of the groups is combined together to arrive at the net HLB value of the product.

The HLB value is calculated from the formula

$$HLB = 7 + \sum H - \sum L$$

Following table lists the contribution of various structural groups.

Table: H and L numbers for representative hydrophilic and lipophilic chemical groups

Hydrophilic groups	H number	Lipophilic groups	L number
$NaSO_4$-	39	-CH-	0.47
KOOC-	21	-CH_2-	0.47
NaOOC-	19	-CH_3	0.47
HOOC-	2.1	=CH-	0.47
HO-	1.9	-CH_2-CH(OH)(CH_3)-	0.11
-O-	1.3	-(CH_2-CH_2-O-)	0.36

Take an example of sodium lauryl sulphate (SLS). The chemical formula of SLS has one sodium sulfato group, eleven methylene groups and one methyl group. Thus we have $\sum H = 39$ and $\sum L = (11*0.47) + 1*0.47 = 5.64$. Inserting these values in the formula, we get the HLB of SLS as 40.36.

For derivatives of fatty acids that have saponifiable groups, HLB value is given by an expression:

$$HLB = 20 \text{ x} \left(1 - \frac{SV}{AV}\right)$$

Where,

SV - The saponification value of the emulsifier

AV - The acid value of the fatty acid used

Thus, for propylene glycol monolaurate, the saponification value is 217 and the acid value of lauric acid is 280.5. Putting these values in the formula, we get the HLB of 4.52.

When the hydrophilic portion is only ethylene oxide, the HLB value is calculated by the equation,

HLB = 20 x W

Where,

W =Weight fraction of ethylene oxide in the surface-active agent

Thus, for 10.5 moles of ethylene oxide condensed on nonylphenol, HLB = 20*[44*10.5]/[218+(44*10.5)] = 13.58

For certain of the non-regular types of ethoxylated esters the saponification value is not readily obtainable and the following formula is used:

$$HLB = \frac{E + P}{S}$$

Where,

E = Percentage ethylene oxide in the molecule

P = Percentage polyalcohol in the molecule

Further, where the hydrophilic portion contains only phenols or mono-alcohols and no polyols, the equation can be simplified to HLB = E/5. The majority of non-ionics fall into this category.

Pigment Volume Concentration (PVC)

The pigment volume concentration refers to the percentage of pigments and fillers relative to the total solids (pigment/filler + resin and additives) in the cured paint film per unit volume.

$$PVC = \frac{V_p}{V_p + V_r} x\ 100$$

$$PVC = \frac{V_p}{V} = \frac{\Sigma V_{pi}}{\Sigma V_i}$$

Where,

V_p = Pigment/filler volume

V_r = Resin and additives volume

V = Total volume

Finish	% PVC
Glossy	17-20
Semi-glossy	25-35
Non-glossy	45 and above
Matt, water based	50+

Critical Pigment Volume Concentration (CPVC)

The point at which there is just enough resin to fill all the void space in the pigment is a compact combination where there are only two phases in exact proportion. The PVC at this position is referred to as critical pigment volume concentration (CPVC)

$$CPVC = \frac{Tapped\ volume\ of\ pigment - volume\ of\ solvent\ used}{Tapped\ volume\ of\ pigment} \text{x } 100$$

It refers to the PVC value in which the amount of resin is not sufficient to fill the inter-particle spaces, and, as a consequence, air can be incorporated into the dry film, causing changes in several properties.

Thus, the CPVC index determines these properties' inflection points. The CPVC varies according to the size and shape of the particles, as it is a direct result of their degree of packaging in the film formation.

The calculation of CPVC is made from the oil absorption (OA) of the pigment/filler, considering a non-flocculated dispersion in oil. OA is expressed in grams of oil per 100 g of pigment.

$$CPVC = \frac{100}{1 + (OA \text{ x } \rho)/K}$$

Where,

ρ = Pigment density

K is 100 times the specific gravity of the binder. For example if linseed oil is used that has the density of 0.935, so K is 93.5

The relationship between PVC and CPVC is known as λ.

λ = PVC/CPVC

This parameter is important to define the type of finish desired in the paint, indicating how close the PVC is to the CPVC.

λ	**Type of finish**
< 0.5	Glossy enamels
0.5 – 0.8	Satin finishing paints
> 0.8	Matte finishing paints
0.5 – 1	Primers
>1	Putty and economic paints

Pigment to Binder Ratio

While the PVC of a coating formulation is a highly meaningful method for comparing paint formulations, the simple ratio of pigment content to the amount of binder is often used in developing formulations. Unless otherwise noted, the basic formula is based on mass. It is especially useful when the densities of the ingredients are not known.

$$Pigment\ to\ binder\ ratio = \frac{M_p}{M_b}$$

Where,

M_p= Mass of pigment

M_b = Mass of binder (resin) in the formula.

Note again that this equation is only useful for the final dry coating, so any solvents present in the formulation must be excluded from the calculation.

$$Pigment\ to\ binder\ ratio = \frac{PVC}{1 + PVC}$$

PVC is expressed as a fraction and not in percentage.

The following methods are used to determine the pigment to binder ratio

(a) Incineration Method

A known weight of sample is taken in a crucible and heated at high temperatures in a furnace so as to evaporate or decompose all the organic material with only inorganic pigment remaining in the crucible. The crucible is then cooled and weighted to get the pigment to binder ratio

Weight of empty crucible = X1

Weight of crucible + paint (before heating) = X2

Weight of crucible + paint (after heating) = X3

$$\% \, pigment = \frac{X3 - X1}{X2 - X1} \, x \, 100$$

$$\% \, Binder = \% \, solid \ content - \% \, pigment$$

From this, the pigment to binder ratio can be calculated.

This incineration method can't be used for paints containing organic pigments, since organic pigment decompose on ignition. Similarly if paint consists of calcium or barium carbonate it may decompose on heating giving error in result.

(b) Centrifuge Method

The sample is diluted with excess of solvent. The resin dissolves in the solvent and the pigment gets separated. Centrifugal force is used to separate the pigment efficiently. This is repeated a couple of times and then the pigment is dried using an oven.

Weight of empty centrifuge tube = X1

Weight of tube with paint = X2

Weight of paint taken = X2 - X1

Weight of the tube after centrifugation and drying = X4

$$\% \, Pigment = \frac{X4 - X1}{X2 - X1}$$

Different solvent mixtures have been suggested for the determination of the pigment content by centrifugation, from which a suitable mixture is to be chosen in accordance with the composition of the coating material. These mixtures are as follows:

Mixtures	Composition	Comments
Mixture A	10 parts by volume of ether 6 parts by volume of benzene 4 parts by volume of methanol 1 part by volume of acetone	It is for oil paints containing the following pigments: All white pigments, ultramarine blue, metallic brown, chrome oxide green, chrome yellows, chrome orange, red oxides, venetian red, yellow iron oxide
Mixture B	5 parts by volume of benzene 4 parts by volume of methanol 1 part by volume of acetone	It is for oil paints containing ferric ferrocyanide blue or chrome green
Mixture C	1 part by volume of toluene 1 part by volume of acetone	It is for nitrocellulose lacquers, petroleum ether for oil paints containing lamp black, umbers, and siennas, toluidine red

Centrifuge separation is carried out for paints which are formulated on organic pigments free from bleeding tendency.

Metal Driers Calculations

The natural drying time of alkyd resins is very long, this is why catalysts, or so-called driers, are added. Generally, these driers are transition metals forming a complex with organic ligands. Driers are often divided into three classes:

1. Catalytic driers are sometimes called top driers, surface driers, or oxidative driers. There are five commercial catalytic driers: Co, Mn, V, Ce and Fe.
2. Crosslinking driers, also called thorough driers, polymerizing driers or coordination driers. They include: Pb, Zr, La, Nd, Al, Bi, Ba and Sr.
3. There is another class of driers that act as auxiliary driers. They include Ca, K, Li and Zn.

$$Metal\ drier\ addition = \frac{W \text{ x } \%\ solid\ content \text{ x } \%\ of\ dosage}{\%\ metal\ content\ of\ drier\ solution} \text{ x } 100$$

Where,

W = Weight of the resin

Contrast Ratio

The contrast ratio is determined from measurements of reflectivity of a paint over the black and white areas of a reflectance chart.

$$C_W = \frac{R_0}{R_W}$$

Where,

R_0 = Reflectance of the coating over the black area of the chart

R_W = Reflectance of the area over the white area of the chart

Hiding Power and Spreading Rate (As per ASTM D 2805)

The hiding power calculation uses Kubelka-Munk theory to determine the film thickness required for hiding, and thus the spreading rate of a given paint.

The reflectivity, R_∞, is then calculated. This is defined as the reflectance of a film thick enough to have the same reflectance over both the black and white areas of the chart.

$$R_\infty = a - (a^2 - 1)^{1/2}$$

Where,

$$a = \frac{1}{2}\left(R_W + \frac{R_0 + W - R_W}{WR_0}\right)$$

R_0= Reflectance of the coating over the black area of the chart

R_W= Reflectance of the area over the white area of the chart

W = Reflectance of the white area of the chart with no coating.

The spreading rate, Hx is calculated:

$$H_X = \frac{AND}{10M}$$

Where,

A = Area of a defined template,

M = Mass of coating contained in the area A

N = Non-volatile content of the coating

D = Density of the paint.

The scattering coefficient, *S*, is calculated next.

$$S = \frac{H_X}{2b}\, ln\left(\frac{1 - R_0R_\infty}{1 - R_0/R_\infty}\right)$$

Where,

$$b = \frac{1}{2}\left(\frac{1}{R_\infty} + R_\infty\right)$$

The hiding power, H_C, is then calculated, usually at a contrast ratio of 0.98:

$H_C = S/P$

Where,

$$P = \frac{1}{2b} ln\left(\frac{U+b}{U-b}\right)$$

$$U = \left[\left(a + \frac{1-C}{1.60C}\right)^2 - \frac{1}{C}\right]^{1/2} + \frac{1-C}{1.60C}$$

$$a = \frac{1}{2}\left(\frac{1}{R_\infty} + R_\infty\right)$$

$$b = \frac{1}{2}\left(\frac{1}{R_\infty} - R_\infty\right)$$

Hiding Power

The hiding power of the pigmented material is a measure of its ability to hide a coloured substrate or differences in the substrate color. The hiding power of the film is determined from the following equation:

$$CR = \frac{L_B}{L_W}$$

Where, CR = Contrast ratio

L_B = Brightness over a black substrate

L_W = Brightness of a white substrate

According to DIN EN 13 300, the coating is considered fully opacifying if the contrast ratio, CR ≥ 0.98.

Refractive Index (ASTM D 542, ISO 489)

It is defined as the ratio of the velocity of light in a vacuum (or air) to its velocity in a transparent medium.

$$Refractive\ index = \frac{\sin of\ angle\ of\ incidence}{\sin of\ angle\ of\ refraction}$$

Mass of binder required for desired DFT:

The following equation relates the quantity of binder required for achieving a certain DFT.

$$m_{binder} = A\ x\ DFT\ x\ \rho_{binder}\ x\ (1 - PVC)$$

Where,

m_{binder} = Mass of binder

A = Area to be covered

DFT = Dry film thickness of the coating

ρ_{binder} = Density of the binder material

PVC = Pigment Volume Concentration of the binder material

Coverage given volume solids and DFT:

The relation between coverage and DFT is as shown below

Unit m^2/kg

$$coverage = \frac{1000\ x\ V_s}{DFT\ x\ \rho}$$

% Volume Solids:

All coatings are made up of resin, pigments, additives and solvent. After the coating is applied, the solvent evaporates, leaving only the dry solids in the form of a continuous coating. The % volume solids is given by:

$$\%\ Volume\ Solids = \frac{Dry\ film\ thickness\ x\ 100}{Wet\ film\ thickness}$$

Relationship between spreading rate and volume solids

If the volume solids is known along with the dry film thickness, the spreading rate can be determined by the following relation

Unit: m^2/litre

$$Spreading\ rate = \frac{\%\ Volume\ solids\ x\ 10}{DFT\ required}$$

Viscosity

Viscosity is an important property for coatings since it dictates its application properties. Viscosity is defined as the property of a fluid which offers resistance to the movement of one layer of fluid over another adjacent layer of the fluid.

Viscosity is also defined as the shear stress required to produce unit rate of shear strain.

Absolute Viscosity (η):

Absolute viscosity also known as dynamic viscosity; is the ratio of shear stress to shear rate. A cone and plate viscometer is the best method for measuring absolute viscosity. Brookfield viscometers also measure absolute viscosity.

Unit: poise (CGS); Pa.s (SI)

Kinematic Viscosity (ν):

Kinematic viscosity is the ratio of the dynamic viscosity to the fluid density. It is commonly determined using a method such as Gardner bubble tubes.

Unit: stokes (CGS).

Efflux Viscosity. Efflux viscosity is determined by measuring the time that it takes for a specified volume of material to exit a cup with a defined orifice size. Common efflux cups are the Ford and Zahn cups.

Unit: seconds

Conversion of the results of one viscosity measurement to another can be challenging. If the liquids are Newtonian, inter-conversion can be carried out through the use of equations that approximately relate the different viscosities to each other.

Absolute viscosity from Kinematic viscosity

$\eta = \rho.\nu$

Where,

ρ = Density of the liquid.

Kinematic viscosity from Ford Cup Viscosity:

No. 2: v = 0.0159t – 0.37

No. 3: v = 0.0233t – 0.16

No. 4: v = 0.036t – 0.16

Other Forms of Viscosity

Relative Viscosity

$$\eta_r = \frac{t.\rho}{t_0.\rho_0}$$

Where,

η_r = Relative viscosity (dimensionless)

t = Run-time of the sample

ρ = Density of the sample

t_0 = Run-time of the pure solvent

ρ_0 = Density of the pure solvent

Specific Viscosity

$$\eta_{sp} = \eta_r - 1$$

Where,

η_r = Relative viscosity (dimensionless)

η_{sp} = Specific viscosity (dimensionless)

Reduced Specific Viscosity

$$\eta_{sp.red} = \frac{\eta_{sp}}{C}$$

Where,

η_{sp} = Specific viscosity (dimensionless)

$\eta_{sp.red}$ = Reduced specific viscosity (cm^3/g)

C = Solution concentration (g/cm^3)

Inherent Viscosity

Dividing the natural logarithm of the relative viscosity by concentration of the solution, obtained the inherent viscosity.

$$\eta_{in} = \frac{Ln\ (\eta_r)}{C}$$

Where,

η_{in} = Inherent viscosity (cm^3/g)

ln (η_r) = Natural logarithm of reduced viscosity (dimensionless)

C = Concentration of the solution (g/cm^3)

The intrinsic viscosity of polymeric solution can be found from the slope of the equation of the line, obtained from the graphic extrapolation to a concentration equal to zero, using the reduced specific viscosity ($\eta_{sp.red}$) and/or the inherent viscosity (η_{in}).

Interrelation of Viscosities

Billmeyer Equation

$$[\eta] = 0.25\left(\frac{\eta_r - 1 + 3.\,ln(\eta_r)}{c}\right)$$

Solomon-Ciuta

$$[\eta] = \frac{\sqrt{2.\,[\eta_{sp} - ln(\eta_r)]}}{c}$$

Deb-Chatterjee

$$[\eta] = \frac{\sqrt[3]{3.\,ln(\eta_r) + 1.5(\eta_{sp})^2 - 3\eta_{sp}}}{c}$$

Where,

$[\eta]$ = intrinsic viscosity

Measuring Instruments

U-tube (Oswald) Viscometer

U-tube (Oswald) Viscometer method gives the measurement of the rate of a flow of a definite volume of liquid through a capillary of definite bore. The time required by the sample to flow from a definite starting point to the end point is recorded. This viscometer is used for the comparison of the viscosities of two liquids and the results are treated in the following way.

If η_1 and η_2 be the viscosities of the two liquids and ρ_1 and ρ_2 be the densities of the two liquids, t_1 and t_2 be the times taken by the two liquids.

Then,

$$\frac{\eta_1}{\eta_2} = \frac{\rho_1 t_1}{\rho_2 t_2}$$

Hence, if one of the viscosities is known e.g. if the distilled water is first used in the apparatus, the viscosity of the second liquid can be calculated from this equation.

Falling Sphere Viscometer

The falling sphere viscometer is based on the measurement of the time taken by a metal sphere to fall through a column of liquid of definite length. It is useful for high-viscosity resin solutions and is based on the principle of Stoke's law:

$$\eta = \frac{2g\,(D - d)r^2}{9V}$$

Where,

η = Viscosity of the liquid

D = Density of sphere

d = Density of the liquid

g = Gravitational constant

V = Terminal velocity of sphere

If a number of spheres of known radius are used and r^2 is plotted against 1/t the result should be a straight line from which η can be obtained.

Gel Content/ % Swelling

The extent of curing is expressed in terms of gel content. The cured material is weighted (initial mass) and then put in test tubes. The material is immersed in MEK. After seven days the material is removed from the solvent, the excess MEK is wiped off and the swollen mass is weighted immediately. The entrapped MEK is removed by heating in a vacuum oven at $85^{O}C$ for 4 hours and the deswollen mass is weighted again. The percent gel content is reported as:

$$\%\,Gel\ content = \frac{Deswollen\ mass}{initial\ mass} \text{x } 100$$

$$\%\,Swelling = \frac{(Swollen\ mass - Deswollen\ mass)}{Deswollen\ mass} \text{x } 100$$

Non Volatile Matter/Solid Content

Solvent is usually added in the coating formulation to ease application and to manage required viscosity. However, it only acts as a medium and the evaporation of this solvent from the material is desirable. Non-volatile matter represents the amount of solid remaining after solvent being evaporated.

To determine the non-volatile matter the known weight of the sample is heated at a temperature so as to evaporate the solvent. Later the final weight is determined.

The amount of non-volatile matter/ solid content is given by following formula

$$Non - volatile\ matter\ or\ solid\ content = \frac{weight\ of\ sample\ after\ heating}{Initial\ weight\ taken} \text{x } 100$$

Solubility Parameter

The net energy required to break the bonds is given by the difference of total energy supplied (latent heat of evaporation, L) and the energy spent on displacing atmosphere (RT, where R is gas constant and T absolute temperature) or L-RT

L-RT is the measure of cohesive energy density. Solubility parameter is defined as square root of cohesive energy density or L-RT/molar volume.

If M is molecular weight of the solvent and D is its density, then the solubility parameter is given by following expression:

$$\delta = \sqrt{\frac{L - RT}{M/D}}$$

Water Vapour Transmission Rate (WVTR) (As Per ASTM E 96)

Water vapour transmission rate, also moisture vapour transmission rate, is a measure of the passage of water vapour through a substance. It is a measure of the permeability for vapour barriers.

$$WVTR = \frac{\Delta m}{At}$$

Where,

A = Area

t = Time

Δm = Change in mass

03

Resin Calculations

03

Resin Calculations

Amine Hydrogen Equivalent Weight

Stoichiometric Ratio of Amine to Use with Epoxy Resin with Known Equivalent Weight

Peroxide Value

Carbonyl Value

Percent Functional Group to Equivalent Weight

Resin Calculations

Acid Value (As per ASTM D974) – *follow the standard to know the titration process*

Acid value is defined as milligrams of potassium hydroxide (KOH) required to neutralise the free carboxylic acid present in one gram of the sample.

$$Acid\ Value = \frac{ml\ of\ KOH\ solution \text{ x } Normality\ of\ KOH \text{ x } 56.1}{weight\ of\ sample\ (solids)\ in\ grams}$$

Unit: mg of KOH/ g of resin.

$$Theoretical\ Acid\ Value = \frac{e_a \text{x } 56100}{Wt}$$

Where,

e_a = Acid equivalents/ functionality

Wt = Weight of the resin solids

$$Carboxyl\ Equivalent\ Weight = \frac{Wt.}{e_a} = \frac{56100}{Acid\ number}$$

Hydroxyl Value (As per ISO 2554 or ASTM D 1957) – *follow the standard to know the titration process*

The hydroxyl value is defined as the number of milligrams of potassium hydroxide (KOH) equivalent to the hydroxyl content of one gram of sample or the number of milligrams of KOH required to neutralise acetic acid capable of combining by acetylation with 1 gram of resin sample.

$$Hydroxyl\ value = \frac{[B - S] \text{ x } normality\ of\ KOH \text{ x } 56.1}{W} - Acid\ value$$

Where,

B = ml of KOH solution required for the reagent blank

S = ml of KOH solution required for the titration of the acetylated sample

W = Weight of the sample used for acetylation

Hydroxyl Number, Hydroxy Equivalent Weight, Percent Hydroxyl

$$Hydroxyl\ number = \frac{e_b \text{ x } 56100}{Wt} = \frac{functionality \text{ x } 56100}{Number\ average\ molecular\ weight}$$

OR

$$Hydroxyl\ number = 33 \text{ x } Percent\ hydroxyl$$

Where, e_b = Number of hydroxyl equivalents

Wt = weight of the resin.

The hydroxyl equivalent weight of a resin is the mass of resin containing one hydroxyl equivalent.

$$Hydroxyl\ Equivalent\ weight = \frac{Wt}{e_b} = \frac{56100}{Hydroxyl\ number}$$

The percent hydroxyl is the weight of hydroxyl in 100 grams of resin

$$Percent\ Hydroxyl = \frac{17}{Hydroxyl\ Equivalent\ weight} \text{ x } 100$$

$$Percent\ Hydroxyl = \frac{functionality \text{ x } 17}{number\ average\ molecular\ weight}$$

Rearranging we get,

$$Hydroxyl\ equivalent\ weight = \frac{17 \text{ x } 100}{Percent\ Hydroxyl}$$

Saponification Value

The saponification value of an oil is the measure of the alkali reactive groups in oil and is expressed as a number of milligrams of KOH per gram of the sample.

OR

It is a measure of the amount of the alkali required to convert a fatty acid or oil into a soap.

OR

The saponification equivalent is defined as the milligrams of the sample which is saponified by 1 mole of KOH.

$$Saponification\ value = \frac{Normality\ of\ KOH \text{ x } 56.1 \text{ x}[B - S]\ in\ ml}{weight\ of\ the\ sample\ in\ grams}$$

Where,

B = ml of KOH for blank titration

S = ml of KOH for sample titration

The saponification equivalent is obtained as follows:

$$Saponification\ equivalent = \frac{56.11}{Saponification\ value}$$

Ester value = Saponification value – Acid value

$$Ester\ content = \frac{Ester\ value}{Saponification\ value}$$

$$\%\ Purity = \frac{Practical\ Saponification\ value}{Theoretical\ Saponification\ value} \text{ x } 100$$

$$Theoretical\ Saponification\ value = \frac{56100 \text{ x } Functionality}{Molecular\ Weight}$$

Isocyanate Content/ Isocyanate Value (As per ASTM D 2572) – *follow the standard to know the titration process*

Percent NCO is the weight percent of isocyanate groups (-N=C=O) present in a resin sample.

% Purity of an isocyanate sample is determined by the following formula:

$$\%\ Purity = \frac{[B - S] \text{x } N \text{ x } E}{1000 \text{ x } W}\ x\ 100$$

Where,

B = ml of acid for blank titration

S = ml of acid for the sample titration

N = Normality of acid used

E = Equivalent weight of the isocyanate being tested

W = Weight in grams of sample used

Percent Isocyanate, Isocyanate Equivalent Weight

$$\%NCO = \frac{42 \text{ x } e_{NCO}}{100 \; grams \; of \; resin}$$

e_{NCO} = Number of isocyanate equivalent/ functionality

$$NCO \; Equivalent \; weight = \frac{4200}{\% \; NCO}$$

Alkyd Resin

Long oil alkyds	Oil content > 55%
Medium oil alkyds	Oil content >45 - 55%
Short oil alkyds	Oil content < 45%

Longer the oil length, more the alkyd reflects oil properties and shorter the oil length, more the properties of the polyester are predominant.

$$Oil \; length = \frac{Weight \; of \; oil \; charged}{Theoretical \; weight \; of \; alkyd} \text{ x } 100$$

To avoid gelation during processing, it is necessary to control the overall functionality of the system, which can be calculated as follows:

$$F_{avg} = \frac{Total\ number\ of\ reactable\ equivalents\ present}{Total\ number\ of\ moles\ present}$$

Alkyd chemist generally prefers to prepare the product with highest molecular weight at lowest acid value without gelation of the batch. To achieve this Carother's theory suggests that the ratio of molar equivalents of hydroxyl to carboxyl groups should approximate to unity.

i.e. $\frac{Equivalents\ of\ OH}{Equivalents\ of\ COOH} = 1$

Practically to avoid gelation and to fulfil other requirements, the above ratio of >1 is taken.

The following table provides excess hydroxyl values that are needed to avoid gelation at low acid values in drying oil alkyds of various oil contents.

% fatty acid content of alkyd	Corresponding % oil content	% excess hydroxyl equivalent	Approximate % excess OH equivalent in the finished alkyd
62	65	0	0
59-62	62-65	5	0-5
57-59	60-62	10	5-10
52.5-57	55-60	18	10-15
48-52.5	50-55	25	15-20
38-48	40-50	30	20-25
29-38	30-40	32	25-30

Epoxide Equivalent Weight

The epoxy value is the number of epoxy groups contained in 100 grams of resin.

$$Epoxy\ equivalent\ weight = \frac{W \text{ x } 1000}{[B - S] \text{x } N}$$

$$Epoxy\ equivalent\ weight = \frac{Weight\ of\ resin}{functionality}$$

Where,

W = Weight of sample in grams

B = ml of alcoholic sodium hydroxide (NaOH) required for blank titration

S = ml of alcoholic NaOH required for sample titration

N = Normality of alcoholic NaOH

Epoxide Equivalent by Titration (as per ASTM D 1652 and D 2074)

$$Epoxy\ equivalent = 1000 \text{ x} \frac{Weight\ of\ sample \text{ x } \%\ non\ volatile\ matter\ (fraction)}{Normality\ of\ HBr \text{ x } volume\ of\ HBr}$$

Epoxy Content

Epoxy content is defined as gram equivalents of epoxy groups in 100 grams of resin.

$$Epoxy\ content = \frac{43 \text{ x } functionality}{Wt\ of\ resin} \text{x } 100$$

Epoxy Equivalent Weight of Resin Mixture

Frequently, epoxy resins are blended, filled, or modified with reactive and non-reactive components. It is therefore necessary to adjust the concentration of the curing agent to cure only the portion of the mix that is reactive; e.g., the resins and any reactive diluent present. This may be simply done by calculating the epoxide equivalent weight of the total mix and then applying previous equation to determine the amount of curing agent to add to 100 parts of formulation.

$$Epoxy\ equivalent\ weight\ of\ mix = \frac{Total\ weight}{\frac{Wt1}{EEW1} + \frac{Wt2}{EEW2} + \frac{Wt3}{EEW3}}$$

Total weight includes all materials, both reactive and non-reactive. The only materials reactive with the curing agent are 1, 2, 3, etc., and are characterised by epoxy ring.

Example

Material	Characteristic	Quantity
Epoxy 1	Average EEW 189	100 parts
Epoxy 2	Average EEW 240	100 parts
Reactive diluent	Average EEW 130	30 parts
Filler		230 parts
Total		460

$$Epoxy\ equivalent\ weight\ of\ mix = \frac{460}{\frac{100}{189} + \frac{100}{240} + \frac{30}{130}} = 391$$

Amine Value (As per ASTM D 2074) – *follow the standard to know the titration process*

It is the number of milligrams of KOH equivalent to the amine alkalinity present in one gram of the sample. The 'amine alkalinity' can be regarded as the number of 'active hydrogen'; they are considered alkaline because the amine group, has a basic character.

Unit: mg of KOH/ g of sample.

$$Amine\ value = \frac{Burette\ reading \text{ x } Normality\ of\ acid \text{ x } 56.1}{Weight\ of\ sample}$$

$$Theoretical\ Amine\ value = \frac{No.of\ amino\ groups\ (functionality) \text{ x } 56100}{Molecular\ weight}$$

Stoichiometric amount of polyamine required to cure given epoxy resin are given as per following formulae:

$$Weight\ of\ polyamine = \frac{Weight\ of\ epoxy \text{ x } Epoxy\ value}{Amine\ value}$$

OR

$$Weight\ of\ polyamine = \frac{56100}{Epoxy\ equivalent\ weight} \text{ x } \frac{Weight\ of\ epoxy\ resin}{Amine\ value}$$

Amine Equivalent Weight

The equivalent weight is the weight of material containing one amine equivalent.

$$Amine\ Equivalent\ weight = \frac{Wt}{functionality} = \frac{56100}{Amine\ value}$$

Amine Hydrogen Equivalent Weight

$$Amine\ hydrogen\ equivalent\ weight = \frac{Molecular\ weight\ of\ amine}{Number\ of\ active\ hydrogens}$$

Stoichiometric Ratio of Amine to Use with Epoxy Resin with Known Equivalent Weight

$$Phr\ of\ amine = \frac{Amine\ hydrogen\ equivalent\ weight \text{ x } 100}{Epoxide\ equivalent\ weight\ of\ resin}$$

Peroxide Value – *follow the standard to know the titration process*

The peroxide value of any sample is the percentage of peroxide oxygen present in the sample.

For example, when benzoyl peroxide is dissolved, the liberated oxygen is used to liberate iodine from potassium iodide. The liberated iodine is titrated against standard sodium thiosulphate.

$$\%\ Peroxide = \frac{V \text{ x } N \text{ x } 242.23 \text{ x } 100}{W \text{ x } 2000}$$

Where,

V = Volume of $Na_2S_2O_3$

N = Normality of $Na_2S_2O_3$

W = Weight of sample

Carbonyl Value (As per ISO 13820) – *follow the standard to know the titration process*

Carbonyl value is defined as the number of milligrams of KOH required to react with one gram of sample.

Unit: mg of KOH/ g of sample

$$Carbonyl\ value = \frac{Volume\ of\ NaOH \text{ x } Normality\ of\ NaOH \text{ x } 56.1}{Weight\ of\ sample}$$

Volume of NaOH = (Main - Blank)

$$\%\ Purity\ of\ Ketone = \frac{Carbonyl\ value \text{ x } molecular\ weight}{56100} \text{ x } 100$$

Percent Functional Group to Equivalent Weight

Functional group content of a resin or polymer is often given in terms of percent functional group (e.g. % isocyanate, % hydroxyl, % epoxy). For stoichiometric calculations, it is often desirable to convert percentage values to an equivalent weight.

Conceptually, percent functional group is the mass of functional group in 100 grams of resin. In terms of equivalents, this is

$$\%\ Functional\ group = \frac{MW_{functional\ group} \text{x} e_{functional\ group}}{W} \text{ x } 100$$

Where,

$MW_{functionalgroup}$ = Molecular weight of the functional group

$e_{functional\ group}$ = Number of equivalents of the functional group

W = Weight of the resin.

Molecular weights of typical functional groups are given in the following table below

Functional group	**$MW_{functional\ group}$**
Isocyanate	42
Hydroxyl	17
Epoxy	43
Carboxylic acid	45

04

Polymer Kinetics & Reaction Rates

Polymer Kinetics & Reaction Rates

Free Radical Chain Polymerisation

To initiate the polymerisation reaction, there must be free-radicals available, which are usually generated by the homolytic decomposition of an initiator. If I is the initiator, $R^{\bullet}$ the free radicals formed and k_d the initiator decomposition rate constant. The rate of initiator decomposition (R_d) is given by:

$$\frac{-d[I]}{dt} = R_d = 2k_d[I]$$

From Arrhenius equation, the temperature dependence of k_d is:

$$k_d = Ae^{\frac{-E}{RT}}$$

Where,

A = Frequency factor

E = Activation energy

R = Gas constant

T = Temperature

From above equations,

$$R_d = 2Ae^{\frac{-E}{RT}}$$

This shows that initiator decomposition increases with both temperature and initiator concentration.

Radicals then attack monomers (M) to generate new radical species.

If M is the monomer and k_i is the initiation rate constant, the rate of initiation R_i is:

$$\frac{-d[M]}{dt} = R_i = k_i[R^{\cdot}][M]$$

Rate of Initiation

Assuming steady initiation, the rate of effective radical formation equals the rate of initiator decomposition:

$$R_i = 2fk_d[I]$$

Where f is the initiator efficiency (typically 0.6-1), representing the fraction of radicals that successfully initiate chain growth.

Rate of Propagation

$$\frac{-d[M]}{dt} = R_p = k_p[M^{\cdot}][M]$$

OR

$$[M^{\cdot}] = \frac{R_p}{k_p[M]}$$

Where,

k_p is the propagation rate constant.

Rate of Termination

$$\frac{-d[M^{\cdot}]}{dt} = R_t = 2k_t[M^{\cdot}]^2$$

Where,

k_t is the termination rate constant.

Case of Steady State

The steady state is when the number of chain growths initiated equals the number of chain growths arrested, the rate of initiation is equal to the rate of termination, i.e., $R_i = R_t$.

It can also be written as:

$$\frac{R_p}{k_p[M]} = \left(\frac{k_d f[I]}{k_t}\right)^{1/2}$$

Leading to:

$$R_p = k_p \frac{k_d^{1/2}}{k_t^{1/2}} (f[I])^{1/2}[M]$$

This relationship shows:

- Rate of polymer formation $\propto$ [monomer]^1
- Rate of polymer formation $\propto$ [initiator]$^{(1/2)}$

Kinetic Chain Length

The kinetic chain length is the average number of monomer molecules consumed by each effective initiator radical. It represents the number of monomer molecules added to a growing chain from its initiation up to termination. The kinetic chain length can be expressed as

$$v = \frac{rate\ of\ propagation}{rate\ of\ initiation} = \frac{rate\ of\ propagation}{rate\ of\ termination}$$

OR

$$v = k_p \frac{k_d^{1/2}}{k_t^{1/2}} (f[I])^{1/2}[M] \frac{1}{2k_d f[I]} = \frac{k_p[M]}{2(k_d k_t f[I])^{1/2}}$$

The degree of polymerisation (D_p) can be correlated to kinetic chain length (ν) as follows:

1. $D_p = 2v$, when termination occurs by coupling
2. $D_p = v$, when termination occurs by disproportionation

Rate of Propagation in Cationic Polymerisation

In cationic polymerisation, the propagation rate is second order with respect to monomer concentration.

$$R_p = \frac{-d[M]}{dt} = k_p[M][M^+] = \frac{k_p k_i}{k_t}[M]^2[H^+]$$

Thus, the propagation rate term contains $[M]^2$ instead of [M].

Rate of Propagation in Anionic Polymerisation

If IA is the initiator, HA is a protonating agent and k_d, k_i, k_p, and k_t are the dissociation, initiation, propagation and termination rate constants,

The propagation rate is given by:

$$R_p = \frac{k_i k_p}{k_t} \frac{[A^+][M]^2}{[HA]}$$

The degree of polymerisation is governed by the following equation.

$$D_p = \frac{k_p}{k_t} \frac{[M]}{[HA]}$$

Further, the expression for the degree of polymerisation depends on whether the initiating species is a monoanion (e.g. Butyl lithium) or a dianion (e.g. sodium complex of naphthalene).

For monoanions (where the chain growth takes place at one end only)

$$D_p = \frac{[M]}{[IA]}$$

For dianions (where the chain growth takes place at both the ends),

$$D_p = 2\frac{[M]}{[IA]}$$

Case of Polycondensation Reactions

Assume that N_A moles of reactant A-A are reacted with N_B moles of reactant B-B and that $N_A < N_B$, giving $r < 1$:

$$\frac{N_A}{N_B} = r$$

Number of reactive groups A to start with = $2N_A$

Number of reactive groups B to start with = $2N_B$

Total number of reactive groups to start with = $2N_A + 2N_B = 2N_A\,[1 + (1/r)]$.

When the extent of reaction is p,

Fraction of groups A consumed = p.

Fraction of groups B consumed = rp

Fraction of groups A remaining unreacted = 1-p

Fraction of groups B remaining unreacted = 1-rp

Total number of groups remaining unreacted = $2N_A\,(1\text{-}p) + 2N_B\,(1\text{-}rp)$.

Now the degree of polymerisation is given by:

$$D_p = \frac{r+1}{r+1-2rp}$$

Special cases:

- When r = 1 (stoichiometric balance)

$$D_p = \frac{1}{1-p}$$

- When p = 1 (full conversion)

$$D_p = \frac{1+r}{1-r}$$

Bifunctional reactants, often, contain some percentage of monofunctional impurities. These monofunctional impurities upset the stoichiometric balance and bring about a decrease in the degree of polymerisation. Assume that N_A moles of A-A, N_B moles of B-B and N_M moles of a monofunctional impurity of the type B are reacted. The stoichiometric imbalance will be represented by:

$$r = \frac{N_A}{N_B + N_M}$$

05

Network Structure & Crosslinking

Network Structure & Crosslinking

Crosslink Density (Flory's Network Theory)

This equation comes from rubber elasticity theory also called as Flory's Network Theory. The assumption includes that the material behaves as an ideal rubbery network and that the deformation is entropic and elastic (above T_g).

$$Crosslink\ density = \frac{E}{3RT}$$

Where,

E = Elastic modulus

R = Gas constant

T = Absolute temperature

Crosslink Density (Based on Molecular Weight between Crosslinks):

This equation comes from network structural theory based on molecular weight between crosslinks.

$$Crosslink\ density = \frac{2\rho}{\phi M_c}$$

Where,

ν = Crosslink density

ρ = Density of crosslinked rubber

M_c = Average molecular weight between cross-linking points

ϕ = 3, 4 for tri- or tetra- functional network junctions respectively.

Affine Network Theory

In the Affine Network Model, the polymer network is treated as consisting of independent random coils whose chain ends are attached to an elastic continuum. This continuum undergoes affine deformation, meaning the macroscopic deformation is uniformly transferred to the molecular scale.

Shear modulus G relates the stress and the deformation as follows:

$$G = \frac{nk_BT}{V} = \nu k_BT = \frac{\rho RT}{M_s}$$

Where,

G = Shear modulus

ν = Number density of network strands; $\nu = n/V$

ρ = Network density

M_s = number average molar mass of the network strands

R = Gas constant

T = Temperature

k_B = Boltzmann Constant

Soluble Branched Rubber Molecule – Determination of Branching (Huggin's Equation):

Huggin's Equation is used to estimate the branching of rubber molecule. Huggins' constant κ' is a function of the polymer/solvent/temperature of the system, but is independent of molecular weight and molecular weight distribution.

It increases with a decrease in solvent power and branching.

- Lower κ′ → good solvent, more linear polymers
- Higher κ′ → poorer solvent quality or greater branching

$$\frac{\eta_{sp}}{C} = [\eta] + \kappa'[\eta]^2 C$$

Where,

C = Concentration of rubber solution

η_{sp} = Specific viscosity

[η] = Intrinsic viscosity

κ' = Huggins' constant

Flory-Rehner Equilibrium Swelling:

The Flory-Rehner Equation is used to calculate the crosslink density of unfilled vulcanised rubber using swelling behaviour

$$\nu = -\frac{1}{2V_s}\frac{\ln(1 - V_r) + V_r + \chi V_r^2}{V_r^{1/3} - \frac{V_r}{2}}$$

Where,

ν = Crosslink density (moles per unit volume)

V_r = Volume fraction of rubber in swollen equilibrium state

V_s = Mole volume of used solvent

χ = Flory-Huggins' polymer-solvent interaction parameter.

For vulcanised rubber containing filler, V_r is corrected as:

$$\frac{V_r}{V_{rf}} = 1 - \{3c[1 - v_r^{1/3}] + V_r - 1\}\frac{\phi}{1-\phi}$$

Where,

V_{rf} = Volume fraction of filled rubber in swollen gel

ϕ = Volume fraction of filler in unswollen filled rubber

c = Filler rubber interaction parameter

Determination of Crosslink Density: Stress-Strain Method

The Mooney-Rivlin equation can be used to calculate the crosslink density of unfilled vulcanised rubber using the stress-strain data:

$$F = 2A_0(\lambda - \lambda^{-2})(C_1 + C_2/\lambda)$$

$$\frac{F}{2A_0(\lambda - \lambda^{-2})} = C_1 + C_2/\lambda$$

Where,

F = Load

A_0 = Initial cross-sectional area

λ = Extension ratio

ρ = Sample density

R = Gas constant

T = Absolute temperature

M_c = Molecular weight between crosslinks

C_1 and C_2 = Mooney constants that define the hyper-elastic behaviour of the rubber-like material. They are obtained by fitting the experimental stress-strain data, often using a Mooney plot.

C_1 is related to the molecular network structure and the shear modulus of the polymer network. It is proportional to the number of molecular strands per unit volume.

C_2 relates to the shear modulus of polymer entanglement. This constant is considered to be independent of the degree of cross-linking and tends to decrease to zero for highly swollen rubbers.

A plot of $F/[2A_0 (\lambda - \lambda^{-2})]$ versus λ^{-1} yields C_1 at the intercept, then:

$$2C_1 = \frac{\rho RT}{M_c}$$

And hence crosslink density can be determined.

Remarks:

- The plot becomes unreliable when $\lambda^{-1} > 0.95$ due to sensitivity to initial length.
- C_1 is typically derived within $0.95 > \lambda^{-1} > 0.5$.
- To ensure equilibrium, strain rate must be low.
- For automated testing, the C_1 remains stable up to ~30 mm/min extension rate for 100 mm length.

Determination of Crosslink Density: AFM (Atomic Force Microscopy) Method

AFM indentation can be used to estimate crosslink density. For a conical tip with 25° half-angle:

$$\frac{F}{d_R^2} = \left(\frac{8}{\pi}\right) G \cot(25)$$

Where,

F = Applied force

d_R = Penetration depth of AFM tip

G = 2AνRT

Where,

A = Constant (1-1.5)

ν = Crosslink density

R = Gas constant

T = Absolute temperature

06

Solubility Parameters and Related Theories

Solubility Parameters and Related Theories

Gibb's Free Energy

The thermodynamics of solubility states that for a liquid molecule to form a stable and spontaneous dissolution of polymer, the overall change in Gibb's free energy (ΔG) should be negative.

$$\Delta G_m = \Delta H_m - T\Delta S_m$$

Where,

ΔG = Change in Gibb's free energy

ΔH_m = Change in enthalpy

T = Absolute temperature

ΔS_m = Change in entropy

	$\Delta H_m < 0$	$\Delta H_m > 0$
$\Delta S_m > 0$	$\Delta G < 0$ System is spontaneous at all temperatures	System is spontaneous at high temperatures, when $T\Delta S_m$ is large
$\Delta S_m < 0$	System is spontaneous at low temperatures, when $T\Delta S_m$ is small	$\Delta G > 0$ System is non spontaneous at all temperatures

Flory Interaction Parameter (χ)

Solubility parameter is a value of relative solvency which varies for different solvent, thus aiding the identification of true solvents for polymers. This equation gives the Flory interaction parameter (χ), which is a dimensionless quantity that represents the interaction energy characteristic of a given solvent-solute pair. This concept facilitates the accurate prediction of complex solubility behaviour, especially in polymers, and it has proven very impactful as it serves to represent many thermodynamic properties of polymer solutions and blends. The likes of solubility, swelling, osmotic pressure, and many other properties can be comprehended in terms of polymer-solvent and polymer-polymer interaction parameters.

Flory-Huggins equation:

$$\frac{\Delta G_m}{RT} = n_s ln\emptyset_s + n_p ln\emptyset_p + \chi\emptyset_s\emptyset_p(n_s + mn_p)$$

$$\emptyset_s = \frac{n_s}{n_s + mn_p}$$

$$\emptyset_p = \frac{mn_p}{n_s + mn_p}$$

Where,

n = Number of moles

ϕ = Volume fraction

m = Ratio of molar volume between polymer and solvent

χ = Flory interaction parameter

ΔG_m = Gibb's free energy

R = Universal gas constant

T = Absolute temperature

Subscripts s and p are for solvent and polymer

Hildebrand Solubility Parameter (δ):

Hildebrand Solubility parameter (δ) is the most simplest and most widely applicable for all the system. The parameter reflects the total van der Waals force as the root of the cohesive energy density of the solvent, which in turn is derived from the heat of vaporization of the pure substance.

$$\delta = \sqrt{\frac{E}{V_m}} = \sqrt{\frac{\Delta H_v - RT}{V_m}}$$

Where,

V_m = Molar volume of the substance in condensed phase

ΔH_V = Heat of vaporization of the pure substance

R = Universal gas constant

T = Temperature

Moreover, Hildebrand and Scott proposed the heat of mixing as:

$$\Delta H_m = \phi_s \phi_p V_{mix}(\delta_s + \delta_p)^2$$

$$\Delta H_m = n_s V_s \phi_p \left(\delta_s - \delta_p\right)^2$$

Where,

ΔH_m = Heat of mixing

n = Moles of solvent

V = Molar volume of solvent

V_{mix} = Volume of the mixture

δ = Hildebrand solubility parameter

ϕ = Volume fraction

Subscripts s and p are for solvent and polymer.

In application, a prediction of solvency can be done by comparing the δ values of the selected polymer and the respective solvent. Usually it is expected that dissolution occurs when the δ values of the solvent is near the polymer. For instance, it can project that polyethylene (δ=16) dissolves in diethyl ether (δ=15.4) and n-hexane (δ=14.9) but not in ethanol (δ=26.0) however, accurate prediction of solubility characteristics can't be achieved by using only Hildebrand solubility parameter. Many specific interactions can affect solubility, especially H-bonds, polymer morphology, crosslinking, size and shape of solvent molecules, as well as other external variables such as temperature and agitation.

Hansen Solubility Parameter (HSP):

Hildebrand's theory was developed with the assumption of non-polar substances, thus, neglecting the association between molecules in polar substances. However, most common solvents and polymers are polar compounds. Moreover, Hildebrand and Scott's expression for heat of mixing doesn't legitimize negative heats of mixing; this has been proven to be incorrect as solubility under such circumstance can be expected.

Hansen proposed to divide the polar portion into a dipole-dipole contribution and hydrogen bonding contribution, in which both could be determined through experimental measures. Therefore, Hansen solubility parameters (HSP) supplemented Hildebrand's concept by constructing a 3D approach to account for dispersive (δ_D), polar (δ_P), and hydrogen bonding (δ_H). The basic governing equation of HSP is that the total cohesive energy must be the summation between the individual energies that contribute to it. By dividing with the molar volume, the square of the total (δ_T) interaction/solubility parameter is described as the sum of the squares of the three accounted interactions.

$$E = E_D + E_P + E_H$$

$$\delta_T^2 = \delta_D^2 + \delta_P^2 + \delta_H^2$$

Similar in concept to the Hildebrand solubility parameter, the application of HSP can be described as a 'likeness' vector between the polymer and solvent. By depicting polymer solubility as a sphere a solubility parameter distance known as R_a was developed, which is the distance between the HSP value of solvent and centre of the polymer solubility sphere.

Solubility Parameter Distance:

$$R_a = \sqrt{4(d_{D1} - d_{D2})^2 + (d_{P1} - d_{P2})^2 + (d_{H1} - d_{H2})^2}$$

Where, subscripts 1 and 2 represents the values for solvent and polymer respectively.

A value known as the polymer interaction radius (R_0) was introduced to determine whether the parameters for the solvent and polymer are within an acceptable range. Good solvents are within R_0, and poor ones are outside it. Moreover, a simple composite affinity parameter called the relative energy distance (RED) is defined as the ratio between R_a and R_0. RED numbers less than 1 is an indication of high affinity and good solvency. Lower affinities and poor solvencies will have progressively higher RED values.

$$RED = \frac{R_a}{R_0}$$

Partition Coefficient (For Small Molecule between Polymer and Solvent):

The partition coefficient describes how a substance (solute) distributes itself between a polymer phase and another phase, such as a liquid or gas, at equilibrium. It is the ratio of the solute's concentration in the polymer to its concentration in the other phase, indicating the polymer's affinity for that substance. This parameter is crucial for understanding solute diffusion, release, and sequestration in polymer-based materials, influencing applications like drug delivery, environmental monitoring, and food packaging.

A high partition coefficient indicates the solute has a greater affinity for the polymer, suggesting it's more lipophilic or less polar.

$$K = \frac{C_{polymer}}{C_{other}}$$

07

Rheology & Viscosity

Shear Stress

Shear Strain

Newton's Law of Viscosity

Power-Law (Ostwald-de Waele)

Viscosity-Temperature (Arrhenius)

Application of Time-Temperature Superposition Principles to Rheology

Storage (Elastic) Modulus

Loss Modulus

Loss Factor / Damping Factor / Tan δ

Capillary Rise Method

Critical Surface Energy (Zisman Plot)

Estimation of Hydrodynamic Radius from Viscosity via Stokes-Einstein (For Small Probes)

Rheology & Viscosity

Viscosity is defined as the property of a fluid which offers resistance to the movement of one layer of fluid over another adjacent layer of the fluid. When two layers of a fluid, a distance 'dy' apart, move one over the other at different velocities, say u and u + du, the viscosity together with relative viscosity causes a shear stress acting between the fluid layers.

Shear Stress:

$$\tau = \frac{F}{A}$$

Shear Strain:

$$\gamma = \frac{du}{dy}$$

Newton's Law of Viscosity:

The Newton's law states that the shear stress is proportional to the shear strain.

$$\tau \propto \gamma$$

$$\tau = \eta\gamma$$

Where,

η = proportionality constant also known as viscosity.

Power-Law (Ostwald-de Waele):

A Power-Law fluid, or the Ostwald-de Waele relationship, is a type of generalized mathematical relationship which approximately describes the behaviour of a real Newtonian and non-Newtonian fluid. Power-law fluids can be subdivided into three different types of fluid based on the value of their flow behaviour index: pseudoplastic, newtonian fluid, and dilatant

The equation is given as:

$$\tau = K\gamma^{n}$$

Where,

τ = Shear stress

γ = Shear strain

K = Flow consistency index ($Pa.s^{n}$)

n = flow behaviour index (dimensionless)

The quantity

$$\mu_{eff} = K\gamma^{n-1}$$

Represents an apparent or effective viscosity as a function of the shear rate. The value of K and n can be obtained from the graph of log (μ_{eff}) and log (γ). The slope line gives the value of n – 1 from which n can be calculated. The intercept at log (γ) = 0 gives the value of log (K).

If $n < 1$; the power-law predicts that the material is pseudoplastic

If n = 1; the power-law predicts that the material is Newtonian fluid

If n > 1; the power-law predicts that the material is dilatent

Viscosity-Temperature (Arrhenius):

The Arrhenius equation for viscosity models how a fluid's resistance to flow (i.e. viscosity) changes with temperature (T) showing it decreases as T increases.

$$\eta = Ae^{\frac{E_a}{RT}}$$

Where,

A = Pre-exponential factor

E_a = Activation energy for flow

R = Gas constant

T = Absolute temperature

η = Viscosity

Application of Time-Temperature Superposition Principles to Rheology:

Polymeric materials, because of their viscoelastic nature, exhibit behaviour during deformation and flow which is both temperature and time (frequency) dependent. For example, when a polymer is subjected to a constant load, the deformation or strain will increase over a period of time. This occurs because the material under load undergoes molecular rearrangement in an attempt to minimize localized stresses. Hence, compliance or modulus measurements performed over a short span result in inaccurate evaluation. This time dependent behaviours suggest to test the material under the actual temperature and time conditions the material will see in the application. This implication, if true, would present real difficulties for the rheologist because the range of temperature and/ or frequencies covered by a specific instrument might not be adequate, or at best might result in extremely long and tedious experiments.

Fortunately, that type of tedious testing is not necessary. Using multi-frequency or accelerated temperature measurements and theoretical time/temperature superposition treatment of the data, limited lab tests (creep, fixed frequency oscillation, or stress relaxation) are sufficient to project long-term properties under a variety of conditions.

The underlying bases for time/temperature super-positioning are (1) that the process involved in molecular relaxation or rearrangements in viscoelastic materials occur at accelerated rates at higher temperatures and (2) that there is a direct equivalency between time (the frequency of measurement) and temperature. Hence, the time over which these processes occur can be reduced by conducting the measurement at elevated temperatures and transposing (shifting) the resultant data to lower temperatures. The result of this shifting is a "master curve" where the material property of interest at a specific end-use temperature can be predicted over a broad time scale.

The amount of shifting along the horizontal (x-axis) in a typical TTS plot required to align the individual experimental data points into the master curve is generally described using one of two common theoretical models.

The first of these models is the Williams-Landel-Ferry (WLF) equation:

$$\log(a_T) = \frac{-C_1(T - T_0)}{C_2 + (T - T_0)}$$

Where, C_1 and C_2 are constants. The values of which depend on the type of polymer and are valid over a limited range of temperature, from T_g to about $T_g + 50^{O}C$. For many amorphous polymers, it has been found that C1 = 17.4 and C2 = 51.6. T_0 is the reference temperature (in K), T is the measurement temperature (in K), and a_T is the shift factor. The WLF equation is typically used to describe the time/temperature behaviour of polymers in the glass transition region. The equation is based on the assumption that, above the glass transition temperature, the fractional free volume increases linearly with respect to temperature. The model also assumes that as the free volume of the material increases, its viscosity rapidly decreases.

The other model commonly used is the Arrhenius equation:

$$\log a_T = \frac{E}{R\,(T - T_0)}$$

Where, E is the activation energy associated with the relaxation, R is the gas constant, T is the measurement temperature, T_0 is the reference temperature, and a_T is the time-based shift factor. The Arrhenius equation is typically used to describe behaviour outside the glass transition region, but has also been used to obtain the activation energy associated with the glass transition.

Storage (Elastic) Modulus:

The storage modulus (G') represents the elastic portion of the viscoelastic behaviour, which quasi describes the solid-state behaviour of the sample. Sometimes, G' is also called elastic shear modulus.

Loss Modulus:

The loss modulus (G”) characterizes the viscous portion of the viscoelastic behaviour, which can be seen as the liquid-state behaviour of the ample. It is also called as viscous shear modulus.

Viscoelastic solids with G’ > G” have a higher storage modulus than loss modulus. This is due to links inside the material, for example chemical bonds or physical-chemical interactions. On the other hand, viscoelastic liquids with G” > G’ have a higher loss modulus than storage modulus. The reason for this is that, in most of these materials, there are no such strong bonds between the individual molecules. The un-crosslinked polymer molecules, for example, are entangled but not chemically crosslinked.

Loss Factor/ Damping Factor/ Tan δ:

The loss factor/ tan δ is calculated as the ratio of loss modulus to storage modulus.

$$Tan\ \delta = \frac{G''}{G'}$$

Where,

G” = Loss modulus

G’ = Storage modulus

For ideally elastic behaviour δ = 0°. There is no viscous portion. Therefore, G” = 0 and with that tan δ = G”/G’ = 0

For ideally viscous behaviour δ = 90°. There is no elastic portion. Therefore, G’ = 0 and thus the value of tan δ approaches infinity.

Capillary Rise Method:

Capillarity is defined as a phenomenon of rise or fall of a liquid surface in a small tube relative to the adjacent general level of liquid when the tube is held vertically in the liquid.

The following equation can be used to determine surface tension of unknown liquid or solution.

$$h = \frac{4\sigma}{\rho g d}$$

Where,

h = Height of rise or fall of level in capillary

σ = Surface tension of the liquid

ρ = Density of the liquid

g = Gravity

d = Diameter of the capillary tube

Critical surface energy (Zisman plot):

The method according to Zisman is a method for investigating the wettability of a solid by determining the critical surface tension with the help of the contact angle. In the Zisman investigation, the cosine of the contact angle θ is plotted against the surface tension of the appropriate liquid. The value of the surface tension extrapolated to cos θ = 1 (contact angle $=0^{O}$) is referred to as the critical surface tension.

Estimation of Hydrodynamic Radius from Viscosity via Stokes-Einstein (For Small Probes):

The equation was introduced by Einstein and is a theory for diffusion coefficients based on molecular-kinetic theory. The approach assumes spherical particles diffusing in a liquid phase consisting of molecules small enough for the phase to be considered as a continuum. The particles on the other hand have to be sufficiently large to ensure that the macroscopic Stokes relation holds. The Stokes equation is strictly justified only at a continuum-scale description

The equation shows that if the temperature and the viscosity of the solvent at a temperature are accurately known, the diffusion coefficient can be converted into the hydrodynamic radius of the solvated particles.

Please note that the Stokes-Einstein equation is generally only valid for dilute solutions and spherical molecules.

$$D = \frac{k_B T}{6\pi\eta r_h}$$

Where,

D = diffusion coefficient

k_B = Boltzmann's constant

T = absolute temperature

η = solvent viscosity; r_h = radius of the particle

08

Mechanical Properties & Analysis

- Tensile Stress
- Tensile Strain
- Flexural Strength
- Poisson's Ratio
- Young's Modulus
- Rigidity Modulus / Shear Modulus
- Bulk Modulus
- Compressibility
- Inter-relation Between Different Modulus
- Compressive Strength
- Fracture Toughness
- Creep Compliance

Mechanical Properties & Analysis

Tensile Stress:

The force applied to produce deformation in a unit area of a test specimen. Stress is a ratio of applied load to the original cross-sectional area of the material.

$$Tensile\ Strength = \frac{Force}{Cross - sectional\ area}$$

$$Tensile\ strength\ at\ yield = \frac{Maximum\ load}{Cross - sectional\ area}$$

$$Tensile\ strength\ at\ break = \frac{Load\ recorded\ at\ break}{Cross - sectional\ area}$$

Tensile Strain:

The ratio of the elongation to the gauge length of the test specimen, or simply stated, change in length per unit of the original length. It is expressed as a dimensionless ratio.

$$\varepsilon = \frac{\Delta L}{L_o}$$

Where,

ε = Tensile strain

ΔL = Change in length

L_o = Original length of the specimen

Flexural Strength:

Flexural strength, also known as bend strength or modulus of rupture, is a material's ability to resist bending forces or loads without fracturing, measuring the maximum stress it can withstand before failure, often determined by tests like the three-point being test where a supported specimen is loaded at its centre.

The flexural stress is defined as the internal stress in a material caused by bending forces, creating tension on one side and compression on the other and is given by:

$$S = \frac{3PL}{2bd^2}$$

Where,

S = Stress

P = Load

L = Length of span

b = Width

d = Thickness of the specimen

Flexural Strain is defined as the fractional change in length when stress is applied on the specimen under test and is given by

$$r = \frac{6Dd}{L^2}$$

Where,

r = strain

L = Length of span

D = Deflection

d = Thickness of the specimen

Poisson's Ratio:

Ratio of lateral stain to axial strain in an axial loaded specimen.

$$v = \frac{Laterial\ strain}{Axial\ strain}$$

Young's Modulus:

The ratio of stress to corresponding strain below the proportional limit of a material. This is also known as Modulus of elasticity. A modulus is a measure of material's stiffness.

$$E = \frac{\sigma}{\varepsilon}$$

Young's modulus is also related to the molecular weight of the crosslinked material by the following equation

$$E = \frac{3\rho RT}{M_c}$$

Where,

E = Young's modulus

ρ = Density of the material

R = gas constant

T = absolute temperature

M_c = molecular weight of polymer between crosslinks

Rigidity Modulus/ Shear Modulus

When a body is loaded within its elastic limit, the ratio of shear stress and shear strain is constant, this is called as rigidity modulus/ Shear modulus.

$$G = \frac{Shear\ stress}{Shear\ strain}$$

Bulk Modulus

When a body is subjected to three mutually perpendicular like stresses of same intensity then the ratio of direct stress and the volumetric strain of the body is

$$K = \frac{Stress}{Volumetric\ strain}$$

Compressibility:

The compressibility is the reciprocal of the bulk modulus of elasticity.

$$Compressibility = \frac{1}{K}$$

Where, K = Bulk modulus

Inter-relation between different modulus

Relation between Young's Modulus and Bulk Modulus

$$E = 3K(1 - 2\nu)$$

Where,

E = Young's modulus

K = Bulk modulus

ν = Poisson's ratio

Relation between Young's Modulus and Shear Modulus

$$E = 2G(1 + \nu)$$

Where,

E = Young's modulus

G = Shear modulus

ν = Poisson's ratio

Relation between Young's Modulus, Bulk Modulus and Shear Modulus

$$E = \frac{9KG}{3K + G}$$

Where,

E = Young's modulus

G = Shear modulus

K = Bulk modulus

Compressive Strength

Compressive strength depends on the stiffness of the material, thus, all of the parameters which affect stiffness, including the effect of fillers, influence compressive strength. The following equation associates compressive strength with other mechanical properties:

$$\sigma_{cc} = KG_R = K\left[\frac{E_\tau}{2(1-\upsilon)}\right]$$

Where,

σ_{cc} = Compressive strength

K = Constant

G_R = Matrix shear modulus

E_τ = Young's modulus

ν = Poisson's ratio

This equation shows that the parameters which affect Young's modulus may affect the compressive strength.

Fracture Toughness:

It is a material property that quantifies its ability to resist crack propagation and failure under applied stress. Fracture toughness is an indication of the amount of stress required to propagate a pre-existing flaw.

The stress intensity factor (K) is a measure of the stress state at the tip of a crack, influenced by the applied stress, crack size, and geometry. A material will fracture when the stress intensity factor reaches or exceeds its fracture toughness (K_{IC}).A fracture has three modes –

Mode I: opening (tensile) mode

Mode II: sliding mode

Mode III: tearing mode.

Of the three modes, Mode I is the most commonly occurring mode of fracture.

The stress intensity factor (for Mode I type of fracture) is given by the following equation.

$$K_I = \sigma\beta\sqrt{\pi a}$$

Where,

K_I = Stress intensity factor

σ = Applied stress

β = Dimensionless correction factor depends on specimen geometry/ geometry factor

a = Crack length

Creep Compliance:

Creep compliance describes a material's tendency to deform over time under a constant load, calculated as the ratio of strain to stress. It is given by the following equation

$$J = \frac{\varepsilon}{\sigma_0}$$

Where,

ε = strain

σ_0 = stress

09

Thermal Properties & Analysis

Heat Capacity (Constant Pressure)

Fourier's Law of Heat Conduction

Thermal Diffusivity

Thermal Conductivity

Coefficient of Linear Thermal Expansion

Thermogravimetric Activation Energy (Kissinger)

Thermal Properties & Analysis

Heat Capacity (Constant Pressure):

Heat capacity or thermal capacity is a physical property of matter, defined as the amount of heat to be supplied to an object to produce a unit change in its temperature.

The amount of heat needed to raise the temperature of unit mass of a material by $1^{O}C$ is called specific heat capacity.

Lower specific heat capacity –Warms up and cools down quickly as it takes much less energy to change its temperature.

Higher specific heat capacity – Warms up and cools down slowly as it takes up much more energy to change its temperature.

Unit: Joule/kelvin (J/K) or Joule/degree Celsius ($J/^{O}C$)

$$C_p = \left(\frac{\partial H}{\partial T}\right)_p$$

Where,

C_p = Heat capacity

∂H= Change in heat

∂T = Change in temperature

Another relation related to heat capacity is as follows:

$$q = mC_p\Delta T$$

Where,

q = Heat energy (J)

m = Mass of the substance (g)

C_p = Specific heat capacity (J/K or J/^OC)

ΔT = Change in temperature (^OC or K)

Fourier's Law of Heat Conduction:

Thermal conductivity is defined as the rate at which heat is transferred by conduction through a unit cross-sectional area of a material when a temperature gradient exists perpendicular to the area.

$$q = -k\Delta T$$

Where,

q = Heat flux

k = Thermal conductivity of the material

ΔT = Temperature gradient

The negative sign indicates that heat flow is in the direction of a decrease in temperature, so heat flows from hotter to colder regions.

Thermal Diffusivity (α):

Thermal diffusivity represents how fast heat diffuses through a material and is defined as the ratio of heat conducted to the heat stored.

Unit: m^2/s

$$\alpha = \frac{Heat\ conducted}{Heat\ stored} = \frac{k}{\rho C_p}$$

Where,

α = Thermal diffusivity

k = Thermal conductivity

ρ = Density

C_p = Specific heat capacity

Together, ρC_p can be considered the volumetric heat capacity

A small value of thermal diffusivity means that heat is mostly absorbed by the material and only a little amount of heat will be conducted further.

Thermal Conductivity:

It is also called as K factor. It is expressed as the quantity of heat that passes through a unit cube of the substance in the given unit of time when the difference in temperature of the two faces is $1^{O}C$. The quantity of heat flow depends up on the thermal conductivity of the material and upon the distance the heat must flow. This relationship is expressed as:

$$K = \frac{Qt}{A(T_1 - T_2)}$$

Where,

K = Thermal conductivity (BHU/in./h/ft^2/F)

Q = Rate of heat flow (BHU/h)

A = Area under the test (in.^{2})

t = Thickness of the specimen (in.)

T_1 = Temperature of hot surface (^{O}F)

T_2 = Temperature of cold surface (^{O}F)

Coefficient of Linear Thermal Expansion

The coefficient of linear thermal expansion is defined as the fractional change in length or volume of a material for a unit change in the temperature.

$$X = \frac{\Delta L}{L_0 \Delta T}$$

Where,

X = Coefficient of linear thermal expansion/^{O}C

ΔL = Change in the length of sample due to heating or cooling

L_0 = Original length of sample at room temperature

ΔT = Temperature difference (^{O}C)

Thermogravimetric Activation Energy (Kissinger):

The Kinssinger method is a popular way of estimating the activation energy of thermally stimulated processes studied by DSC, DTA, and DTG.

Kissinger plot of ln (β/T_p^2) vs $1/T_p$ leading to straight line with slope equal to $-E_a/R$

$$ln\left(\frac{\beta}{T_p^2}\right) = -\frac{E_a}{RT_p} + constant$$

Where,

T_p = Peak temperature

β = Heating rates

E_a = Activation energy

R = Universal gas constant

10

Crystallinity & Morphology

Degree of Crystallinity (by DSC)

Lamellar Thickness (Gibbs-Thomson)

Inter-Planar Spacing (Bragg's Law)

Crystallite Size (Scherrer's Equation)

Crystallinity & Morphology

Degree of Crystallinity (by DSC):

Crystallinity refers to the degree of structural order in a solid.

$$X_c = \frac{\Delta H_m}{\Delta H_m^0} x\, 100$$

Where,

X_c = Degree of crystallinity

ΔH_m = Melt enthalpy, determined as the area under the melt endotherm

ΔH^0_m = Melt enthalpy of a 100% crystalline sample

Lamellar Thickness (Gibbs-Thomson):

Lamelle refers to the platelet-like structure in polymer crystals. This is an important property of the material since it influences polymer's mechanical, thermal and insulating properties. Its thickness can be given by

$$T_m = T_m^0 \left(1 - \frac{2\sigma_e}{\Delta H_f^0 l}\right)$$

T_m = Melting temperature

T^0_m = Melting temperature of bulk crystal

σ_e = Top and bottom fold surface free energy which can be obtained from polymer nucleation theory

l = Lamellar thickness

ΔH^0_f = Heat of fusion per cubic centimetre of the perfect crystal

Inter-Planar Spacing (Bragg's Law):

It is a fundamental principle in XRD that describes the conditions for constructive interference of X- rays reflected form crystalline planes. It states that when an X-ray beam of wavelength λ strikes a crystal lattice with atomic planes spaced by distance d, constructive interference occurs at an angle of incidence θ only if the reflected beams are in phase, a condition met when the path difference is an integer multiple (n) of the wavelength. Measuring these diffraction angles and using Bragg's law allows for the calculation of the inter-planar spacing (d) within a crystal, which in turn reveals the crystal's atomic and molecular structure. The Bragg's law is expressed as:

$$n\lambda = 2d\sin\theta$$

Where,

λ = X-ray wavelength

θ = Angle of incidence

d = Inter-planar distance

n = Integer order of reflection

Scherrer Equation (Crystallite Size):

The Scherrer equation relates the size of crystallites in a solid to the broadening of a peak in a diffraction pattern. It is used to determine the crystal size of the crystallites in a polymer.

$$L = \frac{K\lambda}{\beta cos\theta}$$

Where,

L = Mean size of the crystalline domains

K = Dimensionless shape factor. The shape factor has a typical value of about 0.9 but varies with the actual shape of the crystallite.

λ = X-ray wavelength

β = Line broadening at half the maximum intensity (FWHM),

θ = Bragg's angle

11

Diffusion & Permeability

Fick's First Law (Steady State)

Fick's Second Law (Transient)

Permeability

Diffusion & Permeability

Fick's First Law (Steady State):

The Fick's first law relates the diffusive flux to the gradient of the concentration. It postulates that the flux goes from regions of high concentration to regions of low concentration, with a magnitude that is proportional to the concentration gradient.

OR

In simplistic terms the concept that a solute will move from a region of high concentration to low concentration region across a concentration gradient.

The rate of diffusion is proportional to the surface area and the concentration difference, it is inversely proportional to the thickness of the membrane.

$$J = -D\frac{dC}{dX}$$

Where,

J = Flow rate of any entity (e.g. gas, vapours) that flows through a unit area of a barrier or sample per unit time.

D = Diffusion coefficient

dC/dX = Concentration gradient across the membrane

The diffusion coefficient D is given as

$$D = \frac{k\,T}{6\pi\eta r}$$

Where,

k = Boltzmann constant

T = Absolute temperature

η = Dynamic viscosity of the solvent or medium

r = Radius of the molecule

The above equation is called Stokes-Einstein equation, which relates the diffusion coefficient to temperature, viscosity and particle size. D is inversely proportional to r. The diffusion coefficient also depends on the nature of the polymer.

Fick's Second Law (Transient):

The Fick's second law predicts how diffusion causes the concentration to change with respect to time. It is a partial differential equation which in one dimension reads:

$$\frac{\partial C}{\partial t} = D\frac{\partial^2 C}{\partial x^2}$$

Where,

C = Concentration

t = time

D = Diffusion coefficient

x = Position

Permeability

Permeability is the ability of the material to allow substances to penetrate through its barrier. Permeability coefficient (P) can be defined as the product of the diffusion coefficient D and solubility coefficient S.

$$P = D \, x \, S$$

Where,

D = Diffusion coefficient

S = Solubility coefficient

Permeability depends on membrane thickness, area, and the driving force. The properties of polymer – (i) Tg and (ii) crystallinity also affects permeability. The other factors affecting permeability is the diffusivity which depends on – molecular size of the penetrant and the choice of the polymer.

Unit- Barrer.

12

Surface & Interfacial Properties (Coatings)

Young's Equation (Contact Angle)

Spreading Coefficient

Capillary Number

Washburn Equation (Capillary Rise)

Surface & Interfacial Properties (Coatings)

Young's Equation (Contact Angle):

The contact angle quantifies the wettability of a solid surface by a liquid. It is defined as the angle between a liquid surface and a solid surface where they meet. More specifically, it is the angle between the surface tangent on the liquid-vapour interface and the tangent on the solid-liquid interface at their intersection. Geometrically, it is the angle formed by a liquid at the three-phase boundary point where a liquid, gas and solid intersect.

The Young's Equation describes the balance at the three-phase contact of solid, liquid and gas as:

$$\gamma_{SV} = \gamma_{SL} + \gamma_{LV} \cos\theta$$

Where,

γ_{SV} = Surface tension of the solid i.e. surface free energy

γ_{SL} = Interfacial tension between solid and liquid

γ_{LV} = Surface tension of the liquid.

Spreading Coefficient:

A spreading coefficient is a measure of the wetting behaviour of a liquid on a solid surface. It is defined as the difference between the interfacial free energy of the liquid-solid interface and the sum of the interfacial free energies of the liquid-vapour and solid-vapour interfaces.

Positive spreading coefficient indicates that the liquid tends to wet the solid surface.

Negative spreading coefficient indicates that the liquid doesn't wet the solid surface and instead forms droplets.

$$S = \gamma_s - (\gamma_l + \gamma_{sl})$$

Where,

γ_s = Interfacial free energy of the liquid-solid interface

γ_l = Interfacial free energy of the liquid-vapour interface

γ_{sl} = Interfacial free energy of the solid-vapour interface

Spreading coefficient by measuring the contact angle:

The spreading coefficient for an oil phase displacing water on a solid surface is given as:

$$S_{o/w} = \gamma_{ow}(cos\theta - 1)$$

Where,

γ_{ow} = oil/water interfacial tension

θ = contact angle measured through the oil phase

Capillary Number:

The capillary number is a dimensionless quantity representing the relative effect of viscous drag forces versus the surface tension forces acting across an interface between a liquid and a gas, or between two immiscible liquids.

$$Ca = \frac{Viscous\ forces}{Surface\ tension\ forces} = \frac{\eta U}{\gamma}$$

Where,

η = Dynamic viscosity of the liquid

U = Characteristic velocity

γ = Surface tension or interfacial tension between the two fluids

Washburn Equation (Capillary Rise):

Washburn equation is a mathematical formula that describes the capillary flow of liquid into a porous material. It shows a linear relationship between the square of the penetration height of liquid and time. The equation is used to measure the pore size distribution characteristics of the material and models individual pores as cylinders.

$$l^2 = \frac{\gamma\ r\ cos\theta}{2\eta} t$$

Where,

η = Dynamic viscosity of the liquid

γ = Surface tension of the liquid

l = Penetration distance into the capillary

r = Radius of the capillary

t = time

θ = Contact angle between the penetrating liquid and the solid wall of the capillary tube

13

Optical Properties of Polymers

Beer-Lambert Law

Reflectance, Transmittance and Absorption

Opacity

Optical Properties of Polymers

Beer-Lambert Law:

The Beer-Lambert law is used to calculate the amount of light absorbed by chromophoric groups in polymers. It relate the radiant power in a beam of electromagnetic radiation, usually ordinary light, to the length of the path of the beam in an absorbing medium and to the concentration of the absorbing species, respectively.

$$A = \varepsilon\, c\, l$$

Where,

A = Absorbance, formerly called the optical density

ε = Absorptivity/ molar absorption coefficient/ absorbancy index/ attenuation coefficient formerly called the extinction coefficient

c = Concentration of the absorbing species

l = Length of the beam in the absorbing medium

Reflectance, Transmittance and Absorption:

Reflectance is the amount of flux reflected by a surface, normalized by the amount of flux incident on it. Transmittance is the amount of flux transmitted by a surface, normalized by the amount of flux incident on it. Any flux not reflected or transmitted is absorbed. Thus,

$$R + T + A = 1$$

Where,

R = Reflectance

T = Transmittance

A = Absorption

Opacity:

Opacity is a material's optical property related to the refractive index (n). When a coating is opaque the pigment particles present in a coating scatter and/or absorb light to prevent it from reaching the substrate.

The higher the refractive index of the pigment, the higher the paint opacity it delivers to a coating. The pigment diffuses/reflects light before it can reach the underlying substrate. This makes the coating film appear opaque

$$Opacity = 1 - \frac{T_{back}}{T_{front}}$$

T_{back}= Represents the light transmitted through the material when the light source is at the back

T_{front} = Represents the light transmitted through the material when the light source is at the front

Opacity can also be described as:

$$K_v = \rho\ x\ ln\left(\frac{I_x}{I_0}\right)$$

Where,

K_v = Opacity of medium when a beam of light of frequency v passed through it

P = Density of medium

x = Distance the light has travelled through the medium

I_x = Intensity of light remaining at distance x

I_0 = Original intensity of light at x = 0

14

Electrical Properties of Polymers

Volume Resistivity

Conductivity

Surface Resistivity

Dielectric Strength

Dielectric Constant (Relative Permittivity)

Capacitance

Percolation Threshold (Conductive Fillers)

Joule Heating

Dissipation Factor

Power Factor

Shielding Effectiveness of Measurement

Electrical Properties of Polymers

Volume Resistivity (ρ):

Volume or bulk resistivity of an isotropic material is the resistance, as determined in accord with Ohm's law between opposite faces of a unit cube. The resistance R between opposite ends of a block of uniform cross sectional area A and length L is given by

$$R = \frac{\rho L}{A}$$

Alternatively, volume resistivity may be expressed by:

$$E = \rho J$$

Where, E and J represent the electric field and current density respectively at any point in the material.

$$\rho = \frac{AV}{It}$$

Where, V = Applied voltage (volts)

I = current (ampere)

A = area (cm^2)

t = thickness (cm)

Unit: Ω m

Conductivity:

Conductivity is reciprocal of volume resistivity,

$$Conductivity = \frac{1}{\rho_v}$$

Unit of conductance: Siemens (S)

Surface Resistivity (σ):

Surface resistivity is the resistance between opposite edges of a square. The resistance across a square is independent of the size of the square. The conducting surface must be in reality be a layer with a finite thickness t, and only an effective surface resistivity which is related to the true volume resistivity of the layer by

$$\sigma = \frac{\rho}{t}$$

Unit: Ω

$$\sigma = R\frac{W}{L}$$

$$\sigma = \frac{PV}{Ig}$$

Where,
P = Effective perimeter of the guarded electrode
g = Distance between the electrodes.

Dielectric Strength (ASTM D 149, IEC 243-1):

The dielectric strength of an insulating material is defined as the maximum voltage required to produce a dielectric breakdown. Dielectric strength is expressed in volts per unit of thickness. The factor affecting the dielectric strength of the material includes specimen thickness, temperature, humidity, electrode geometry (electrode area and base material composition), time, mechanical stress and processing defects.

$$Dielectric\ Strength = \frac{Breakdown\ voltage\ (V)}{Thickness\ (mm)}$$

Dielectric Constant (relative permittivity) (ASTM D2520, ASTM D150, ASTM D1673 and IEC 62631-2-1:2018: Part 2-1):

The dielectric constant in polymers measures their ability to store electrical energy in an electric field relative to vacuum. This property is crucial in electronics applications because it directly affects signal propagation speed, impedance characteristics, and capacitance values in circuit components.

Polymers with low dielectric constants are highly valued in high-frequency applications like telecommunications and computing because they minimize signal delay and cross-talk between conductors. The dielectric constant also influences a polymer's compatibility with specific electronic applications-materials with higher values are better for capacitors and energy storage, while lower values are preferred for insulation and circuit boards. Dielectric constant is affected by temperature, frequency, voltage, structure and morphology, and moisture absorption.

The dielectric constant is defined as the ratio of the capacitance of the material to that of the vacuum.

$$\kappa\ or\ \varepsilon_r = \frac{C}{C_0} = \frac{\varepsilon}{\varepsilon_0}$$

$$C_0 = \frac{\varepsilon_0 A}{T}$$

Where,

K or ε_r = dielectric constant of the material

C = Capacitance using the material as the dielectric capacitor.

C_0 = Capacitance using vacuum as the dielectric

ε = permittivity of the substance

ε_0 = permittivity of free space (8.85×10^{-12} F/m i.e., Farad per metre)

A = area of the plate/ sample cross sectional area

T = Thickness of the sample

Dielectric constant is also called relative permittivity (ε_r). This is because it is measured relatively from the permittivity of free space (ε_0).

Capacitance:

The capacitance of a circuit element is a measure of its ability to hold charges. It is dependent on the geometry and the materials it is made of. Capacitance of conducting polymers originates from charges stored in atomistic Stern layers formed by counter ions and doped polymeric chains. The Stern layers have a purely electrostatic origin, since the counter ions do not form any bonds with the atoms of the polymeric chains, and no charge transfer between the counter ions and conducting polymer takes place.

$$C = \frac{Q}{V}$$

Where,

C = capacitance of the element

Q = magnitude of the charge held on the element

V = Potential difference across the circuit element.

For a parallel plate capacitor, the capacitance is given by the formula:

$$C = \varepsilon \frac{A}{d} = \frac{\kappa \varepsilon_0 A}{d}$$

Where,

A = Area of each plate

d = Distance between the plates

Ɛ = Permittivity of the material between the plates of the parallel capacitor.

ε_0 = Permittivity of free space (8.85×10^{-12} F/m)

κ = Relative permittivity of the dielectric material between the plates

Unit: Farad (F)

Percolation Threshold (Conductive Fillers):

The percolation threshold is the lowest concentration of conductive filler at which insulating material is converted to conductive material. This means that percolation threshold is the lowest concentration of filler at which electrical pathway is formed throughout a sample. The shape, size of particle, size and geometry of aggregate, aspect ratio and dispersion of filler determine potential for formation of pathway.

Percolation threshold can be estimated from the following equation.

$$\sigma = \sigma_0 \left(\frac{\phi - \phi_c}{1 - \phi_c}\right)^{\mu}$$

Where,

σ = DC conductivity

σ_0 = DC conductivity above the percolation threshold

Φ = Volume fraction of filler loading

Φ_C = Volume fraction at percolation or critical volume fraction

μ = Power law exponent

Joule Heating:

Joule heating, also referred to as Ohmic heating or resistive heating, based on the law of conservation of energy, is a phenomenon that occurs when a voltage difference is applied across a conductive material such that electrical energy is consumed in overcoming the resistance of the material causing this energy to be generated in the form of heat. This heating is referred to as Joule's effect.

The amount of heat produced by the material is proportional to the square of the current, the resistance of the circuit, and the duration for which the current flows through the circuit.

$$Q = I^2Rt$$

Where,

Q = Heat produced by the material in Joules

I = Electrical current flowing through the conductor in amperes

R = Electrical resistance in Ohms

t = Elapsed time in seconds

Dissipation Factor (ASTM D2520, ASTM D150 and IEC 62631-2-1):

Dissipation factor is defined as the reciprocal of the ratio between the insulating materials' capacitive reactance to its resistance at a specified frequency. In other words, it is defined as the ratio between the permittivity and the conductivity of an electrically insulating material. The factors like frequency, temperature, humidity, weather conditions affect dissipation factor of any material.

The lower dissipation values (typically below 0.001) indicate superior electrical performance with minimal energy loss and heat generation. This property becomes particularly critical where signal integrity and thermal management are paramount concerns. The dissipation factor varies with frequency, temperature, and humidity, requiring careful consideration of operating conditions while material selection.

It measures the electrical energy absorbed and lost when an electrical current is applied to the material. Most of the absorbed energy is dissipated as heat. It is a dimensionless quantity. It is also called as loss tangent, tangent of the loss angle, or tan $\boldsymbol{\partial}$.

$$\tan\delta = \frac{I_R}{I_C} = \frac{1}{\omega RC}$$

Where,

$\boldsymbol{\partial}$ = phase angle between the voltage and current waveforms in the insulator

I_R = resistive current

I_C = capacitive current

R = voltage-dependent resistance

Ω = angular frequency of the AC voltage applied to the insulator

C = capacitance of the insulator

Power Factor:

The power factor of an insulator is defined as the ratio of power dissipated in watts to total charging volt-amperes. It is also the cosine of the angle between the voltage applied and the resulting current.

$$Power\ factor = \frac{\tan\delta}{\sqrt{1 + tan^2\delta}}$$

Where,

tan $\boldsymbol{\partial}$ is the dissipation factor.

Shielding Effectiveness of Measurement

Electromagnetic energy is most effectively measured in terms of attenuation. Attenuation is the amount by which the intensity of an electromagnetic signal is reduced by the introduction of a shielding medium. The attenuation is measured in decibels (dB) and is the ratio of the field strength without the shield to the field strength with the shield. Shielding effectiveness measured in decibels is a logarithmic scale, which means that an attenuation measurement of 50 dB is 10 times more effective than a measurement of 40 dB.

$$Total\ electromagnetic\ shielding\ effectivness = 10\log Pi/Pt$$

Where,

Pi = Incident power in watts/m^2

Pt = Transmitted power in watts/m^2

	Shielding Attenuation Levels
0-10 dB	Very little shielding
10-30 dB	Minimum range for meaningful shielding
30-60 dB	Average shielding
60-90 dB	Good shielding
90-120 dB	Excellent shielding
Over 120 dB	Limit of present “state of the art”

15

Flame Properties of Polymers

Oxygen Index Test

Surface Burning Characteristics of Materials- Surface Flammability of Materials Using a Radiant Heat Energy Source

Flame Properties of Polymers

Oxygen Index Test (ASTM D 2863, ISO 4589)

Oxygen index is defined as the minimum concentration of oxygen, expressed as volume percent, in a mixture of oxygen and nitrogen that will just support flaming combustion of a material initially at room temperature under specified conditions. The oxygen index test is considered one of the most useful flammability tests because it allows one to precisely rate the materials on a numerical basis and simplifies the selection of plastics in terms of flammability. The test results are affected by the thickness of specimen, presence of fillers and flame retardants.

$$Oxygen\ Index\ Percent = \frac{(100\ x\ O_2)}{(O_2 +\ N_2)}$$

Where,

O_2 = Volumetric flow of oxygen at the concentration determined

N_2 = Volumetric flow of nitrogen

Surface Burning Characteristics of Materials – Surface Flammability of Materials Using a Radiant Heat Energy Source (ASTM E 162)

This test, also known as the radiant panel test, is one of the most widely used laboratory-scale flame-spread tests. Although not recommended for use as a basis of ratings for building code, the test does provide a basis for measuring and comparing the surface flammability of materials when exposed to a prescribed level of radiant heat energy.

The flame-spread index of a specimen is determined by measuring the rate of burning down of the specimen and the heat rise associated with the burning in the stack above the burning specimen.

$$I_s = F_s \, x \, Q$$

Where,

I_s = flame spread index

F_s = flame spread factor

Q = heat evaluation factor.

16

Important Relations in Terms of Fillers

Important relations in Terms of Fillers

Particle Size Distribution:

Smaller particle size with narrow distribution always helps to better reinforce the binder as it leads to better uniformity.

A mathematical form of data presentation is sometimes more convenient than graphical representation of particle size distribution. Granulometry in number (L_n) and in weight (L_w) is calculated from the following equations:

$$L_n = \frac{\Sigma d_i n_i}{\Sigma n_i}$$

$$L_w = \frac{\Sigma d_i^2 n_i}{\Sigma d_i n_i}$$

Where,

d_i = Particle diameter

n_i = Number of particles

The results are either expressed as a ratio- L_w/L_n or a dispersity factor is calculated

$$D = \frac{L_w - L_n}{L_n}$$

Effect of Particle Size on Optical Properties

Fillers may also play the role of pigment, and, when they do, the particle size distribution play an important role. The tint strength and opacity depend on the particle size.

$$d_{opt} \sim \frac{\lambda}{16(n_p - n_B)} \, [nm]$$

Where,

d_{opt} = Optimum particle diameter

λ = Wavelength of the incident light

n_p = Refractive index of pigment

n_B = Refractive index of matrix

According to this relation there is a direct interdependence between scattering power and particle diameter. This equation suggests that pigment having different particle size distributions may have different scattering properties in terms of hiding and opacity. The particle diameter also affects the scattering of blue, green and red light. Changes to the particle size distribution will change the undertone of the pigment allowing a system to be tailored to the requirements. Certain grades may be capable of providing optical brightening or mask the yellow color.

Permeability and Barrier Properties

The effect of filler reinforcement on permeability and barrier properties of the prepared composite is correlated by the following equation

$$\frac{P_c}{P_p} = \frac{\phi_f}{1 + \left(\frac{W}{2T}\right)\phi_p}$$

Where,

P_c = permeability of composite

P_p = permeability of unfilled polymer

Φ_p = volume fraction of polymer

Φ_f = volume fraction of filler

W = particle width

T = particle thickness

Guth-Gold Equation:

The equation shows that the degree of reinforcement increases with increasing filler concentration.

$$\frac{E_f}{E_g} = 1 + 13\phi + 14.3\phi^2$$

Where,

E_f = Storage modulus of filled system

E_g = Storage modulus of unfilled rubber

Φ = Volume fraction of filler

Effect of Fillers on Tensile Strength of the Filled Materials

A general equation describes the effect of the volume fraction of a filler on tensile strength:

$$\sigma_c = \sigma_p\left(1 - a\phi_f^b + c\phi_f^d\right)$$

Where,

σ_C = Tensile strength of composite

σ_p = Tensile strength of polymer matrix

Φ_f = Volume fraction of filler

a, b, c, d = Constants

Without knowing the values of these coefficients, it is not possible to predict if tensile strength of the composite increases or decreases as the volume fraction of the filler increases. It is also obvious from the form of the equation that constants can be selected to describe certain features of the fillers behaviour. For example, constant “a” is usually related to stress concentration. In composites, in which the filler has very poor adhesion a = 1.21 or a =1.23 for non-spherical particles.

The constant “b” is usually assigned to arbitrary value of 0.67.

Constants “c” and “d” related to the effect of particle size. The smaller the particle size, the larger are the values of these constants. When the values of these four constants are known or approximated, it makes it possible to predict the tensile strength of various composites.

Many modifications of the above equation or its parameters (constants) are used to explain experimental data.

For low concentrations of fillers, the Einstein equation usually fits experimental data

$$\sigma_c = \sigma_p\left(1 + a\phi_f^b\right)$$

In the Einstein equation, b = 1 for spherical particles at low concentrations and "a" depends on the adhesion between the matrix and the filler. This equation predicts that the addition of filler increases tensile strength which was found to be not always the case, so this equation has been modified by various researchers. The Nicolais and Narkis equation is a common modification in which a = 1.21 and b = 2/3

A modified Nielsen model is another frequently used equation especially in the form proposed by Nicolais and Narkis.

$$\sigma_c = \sigma_p \frac{1 - \phi_f}{1 + 2.5\phi_f} \exp\left(B\phi_f\right)$$

In this equation "B" is a parameter characterizing the interaction.

Effect of Fillers on Elastic Modulus of the Filled Materials

Elastic modulus (or Young's modulus) is frequently used to characterize filled systems. Einstein's viscosity equation modified by Guth and Gold:

$$E = E_0(1 + 2.5\phi + 14.1\phi^2)$$

Predicts that elastic modulus, E, increases as the filler concentration, Φ, increases.

Effect of Fillers on the Flow of the Material (Fedor's Equation):

The following equation relates the effect of filler on the viscosity of prepared reinforced matrix. Viscosity increase and, therefore, flow decrease depend on filler concentration and the packing factor which is related to filler volume.

$$\frac{\eta}{\eta_p} = \left(\frac{1.25\emptyset}{F - \emptyset}\right)^2$$

Where,

η = Viscosity of filled system

η_p = Viscosity of matrix

Φ = Filler fraction

F = Packing factor

Complex Viscosity:

Complex viscosity is obtained from dynamic mechanical studies. It is calculated from the following equation:

$$\eta^* = \sqrt{\left(\frac{G'}{\omega}\right)^2 + \left(\frac{G''}{\omega}\right)^2}$$

Where,

G' = Storage modulus

G" = Loss modulus

ω = Frequency

Complex viscosity measurement is very useful in estimation of steady shear viscosity in ranges in which shear viscosity cannot be measured but can be obtained from complex viscosity data using Cox-Merz rule:

$$|\eta^*(\omega)| = |\eta\gamma^{\cdot}|_{\gamma^{\cdot}-\omega}$$

Where,

η = Steady shear viscosity

$\gamma^{\cdot}$ = Shear rate

This relationship is especially used in studies of polymer melts (including fillers).

17

Polymer Processing & Mixing

Residence Time in Extruder

Weissenberg Number

Die Swell

Polymer Processing & Mixing

Residence Time in Extruder:

$$t_r = \frac{V_{barrel}}{Q}$$

Weissenberg Number (W_i):

It is a dimensionless number used to estimate the viscoelastic character of a flow. It is the product of the relaxation time and a typical shear rate.

$$W_i = \lambda \gamma$$

Where,

λ = Relaxation time

γ = Shear rate

When W_i is low, generalized Newtonian models are sufficient to describe the flow

At higher values of W_i, viscoelastic models are required to characterize memory effects.

Die Swell

High viscous elastic fluids, such as polymer melts, exhibit an increase in the cross-section diameter on emerging from a capillary. This phenomenon is often referred to as "Barrus effect" or "die swell" and is usually described quantitatively in terms of the swelling ratio.

Where,

$$Die\ swell = \frac{d}{D}$$

d = Extrudate diameter

D = Capillary die diameter

18

Environmental Aging & Degradation

Arrhenius Accelerated Ageing Time Equivalence

UV Dose

Oxidation Induction Time

Carbonyl Index

Environmental Aging & Degradation

Arrhenius Accelerated Ageing Time Equivalence:

The formula is used to relate the time it takes for a product or material to age under real-world conditions to the time it takes to age under accelerated conditions.

$$t_{real} = t_{acc} e^{\frac{-E_a}{R}\left(\frac{1}{T_{real}} - \frac{1}{T_{acc}}\right)}$$

Where,

t_{real} = Time at real-world conditions

t_{acc} = Time at accelerated conditions

E_a = Activation energy of the degradation process

R = Ideal gas constant

T_{real} = Absolute temperature (in Kelvin) at real-world conditions

T_{acc}= Absolute temperature (in Kelvin) at accelerated conditions

UV Dose:

It denotes the amount of energy received by the material. It is given by the following equation

$$Dose = I \ x \ t$$

Where,

I = Intensity of UV (mW/cm^2)

t = Time (seconds)

Oxidation Induction Time (OIT) (ASTM D3895/ EN 728/ ISO 11357):

Many organic compounds tend to undergo oxidation reaction in the presence of oxygen, even at low temperatures. However some classes of substance, when subjected to isothermal conditions, exhibit a period of induction during which no reaction with oxygen occurs. This time period is called the oxidation induction time or oxidative induction time. After this period, a reaction with oxygen occurs at an increasing rate. OIT is frequently used to study the stability of materials like polyolefins, lubricating greases and oils, edible oils and fats, etc. in process development and quality control. The measurement is carried out in an oxygen atmosphere at a temperature that is sufficiently high to ensure that decomposition begins within a reasonable time. OIT is performed under accelerated conditions to shorten the measurement duration for practical feasibility.

OIT is the time from the initial exposure to oxygen to the onset of exothermic decomposition at the isothermal temperature of the test. It gives a relative measure of the stability of the test material toward oxidation. The longer the OIT, the more stable the material.

OIT measurements are commonly performed isothermally by DSC in an oxidative atmosphere. The instrument measures the heat produced in the oxidation reaction and detects the onset of oxidation accordingly. The sample is inserted in an open crucible into the instrument and then rapidly heated to the isothermal test temperature, with nitrogen as a purge gas. Once the sample temperature is stabilized, the purge gas is switched to oxygen. The onset is evaluated as the point of intersection of the tangent to the exothermic reaction curve with the extrapolated DSC baseline.

The activation energy of oxidation can be determined by Arrhenius equation as follows:

$$\ln t_{OIT} = \ln A' + \frac{E_a}{RT}$$

Where,

t_{OIT} = oxygen inhibition time

A = Frequency factor

E_a = Activation energy

R = Universal gas constant

T = Absolute temperature

Carbonyl Index (CI)

Carbonyl index can be used as a tool to detect and monitor the degradation in polymers. Changes in elongation at break and carbonyl index during degradation showed a strong correlation with the average molecular weight of the polymers. The test is carried out by FTIR which detects the changes in the carbonyl group (C=O) at 1746-1510 cm^{-1}. The technique is most relevant to polyolefins as the presence of carbonyl groups indicate that the polymer has been degraded and facilitates further degradation due to the photolability (unstable near light) of carbonyl groups.

$$Carbonyl\ Index = \frac{Carbonyl\ bond\ absorption\ peak\ area\ (1850 - 1650cm^{-1})}{Reference\ bond\ absorption\ peak\ area\ (1500 - 1420cm^{-1})}$$

19

Coating Defects & Remedies

This chapter covers all the possible defects that may occur in the coating along with its causes and remedies.

Coating Defects and Remedies

Defects	Causes	Remedies
Faults that develop in the paint during manufacture and storage		
Settling	Large differences in specific gravity between pigment and binder, very large particle size of pigment, improper wetting and dispersing agent, inadequate thixotropy of the paint and very low viscosity of the medium.	Reduce mass of the particles by reducing particle size or using lower-density pigments. Increase viscosity by using higher molecular weight resin, a different solvent combination, higher resin solids or adding a thixotropic agent.
Medium separation	The top part of the contents become totally devoid of the pigments/fillers, but there is no hard settling. Caused by inadequate or incompatible wetting and dispersing agents.	Use compatible and sufficient quantity of wetting and dispersing agents.
Thickening of the medium	In lacquer systems, this occurs due to inadequate solvation of the resin. If the resin is not fully solvated, the true solvent continue to penetrate the resin which increases the viscosity.	It is advisable that the resin to be heated to about 60-80°C for a couple of hours so as to get complete penetration of the solvent leading to stabilized viscosity.

	In the case of drying oils and alkyds, thickening is caused by various parameters such as polymerisation, reaction of carboxylic acid groups with alkaline pigments such as zinc oxide or lead oxide.	Avoid reactive pigment combinations.
Skinning	Drying of the paint surface due to air in the container headspace.	Use air-tight containers with minimal headspace. Add anti-skinning agents (e.g. methyl ethyl ketoxime). Care must be taken not to add excess of its quantity as it may completely retard drying process.
Darkening of clear varnishes/ light coloured paints	May occur due to corrosion of containers.	Use tin plated or plastic containers. Add adequate quantity of corrosion inhibitor (e.g. amine based inhibitors).
Gassing	In solvent-based systems: reaction of moisture with zinc or aluminium pigments produces hydrogen gas. In water-based systems: microbial contamination causes gas and odour.	Resins, extenders and solvents must be dried completely. Desiccants such as lime to molecular sieves should be used.

		Use biocides for in-can protection as well as film protection.
Faults developed during application of paints		
Blushing	A milky white haze or mist formed on the surface of a film. Moisture condensing on, and being trapped in the wet film. this may be due to • Spraying during the cold, wet or humid weather. • Use of too fast or poor quality thinner. • Compressed air pressure onto the film to speed up solvent release. • Draughty paint shop, or inadequate heating and/or air movement.	If possible, avoid spraying air drying coating during rainy or exceptionally cold or humid weather. If this is unavoidable, a non-bloom correct grade thinner should be used. Reduce compressed air pressure to minimise the cooling effect. Allow solvent release naturally.
Spitting	Improper adjustment of the spray gun causes non uniform discharge of the coating to the substrate when a large quantity of the paint is suddenly thrown to the substrate, the coated material can't	Adjust spray gun pressure and nozzle settings. Clean nozzle regularly.

	flow and level adequately and shows spitting marks.	
Threading, cob webbing and dry spray	Too fast evaporating solvent that evaporates before the coating sets over the surface, high surrounding temperature, windy conditions and improper adjustment of the spray gun causes these defects. Too high resin molecular weight leads to formation of threads and cobweb.	Use slower-evaporating solvents. Adjust gun settings and ambient conditions. Formulate with optimal resin molecular weight.
Improper temperature in the oven	Uneven oven temperature causes non-uniform curing and performance variations.	Inspect the uniformity of temperature in the oven. If the oven is large, adequate provision must be made to obtain adequate air circulation with uniform air distribution.
Too low thickness of the coating film	This happens if coating is excessively diluted.	Optimum dilution should be studied.
Sagging	Sagging occurs due to improper thixotropy. It is expected that as the	Add optimum amount of thixotropic agent.

	solvent evaporates, the viscosity of the medium increases enough to allow flow and levelling but prevent sagging. If the quantity of thixotropic agent is not adequate, sagging may occur.	
Pin holes and cratering	Poor wetting of the substrate causes pin holes and cratering. Poor wetting may be due to poor surface preparation, presence of oil/impurities on surface, large excess of silicon oil as an additive for improving gloss or as defoamer.	Use proper surface preparation techniques- sanding, cleaning (free from oil, dirt, etc.) and substrate condition, wetting agents, etc. Optimise other additives that could affect wetting
Flocculation	Inadequate wetting and dispersing agent may cause flocculation of particles.	Use optimum levels of compatible wetting and dispersing agents
Excessive drying time for air-drying paints	This may be due to inadequate or wrong combination of driers, high humidity of the surrounding.	Adjust drier types and levels. Cobalt for surface drying; lead/zirconium for through drying
Silking, grained appearance	Silking occurs when the resin tends to crystallize out and the crystalline portion is not fully soluble in the solvent employed.	Use compatible resins Store at appropriate temperatures

	This can also occur when the coating material is stored at low temperature for too long. Grained appearance is caused by complete precipitation of resin, that doesn't re-dissolve. Sometimes precipitation of lead driers due to prolonged storage or low temperature storage also leads to silking or grained appearance.	Use proper solvent blend
Wrinkling	Wrinkling is caused by differential drying rates, particularly when surface drying is faster than through drying. Accelerated surface drying is caused by excess of cobalt drier.	Apply thinner coats or add more through drier (lead or zirconium octoate). Add some anti-oxidant (substituted phenols) or dilution of the coating.
Tackiness/ after tack	Insufficient drier	Adjust drier content. Use balanced drier package (surface + through).
Loss of gloss on drying or stoving	In case of paint, loss of gloss may be due to poor formulation (high PVC), poor grinding and dispersing improper solvent combination, improper drying	Identify the problem. PVC and solvent combination would require reformulation whereas poor grinding may require

	conditions such as windy or dusty environment.	reprocessing. Optimize drying conditions and solvent system.
Orange peel effect	• deviating spray pressure, spray viscosity, spray technique, processing temperature • Unsuitable solvent combination • scarred underground • unsuitable spray gun (nozzle) • too fast drying • poorly run primer coating • absorbent substrate • incompatible raw materials • incorrect processing technique • No optimal layer thickness	• Observe processing instructions according to product information • proper surface preparation • suitable spray gun (nozzle) • Use prescribed dilution • Levelling additives help with correction
Pitting, cratering, blister formation	Ineffective surface wetting of the surface causes pitting and cratering.	Remedies for pitting and cratering are better surface preparation, removal of oil droplets from

	Entrapment of air in form of trapped bubbles causes formation of blisters. If the blister breaks at too late at stage when solvent evaporation is completed and resin can no longer flow, it leads to defects on the film surface.	the compressed air, better wetting agent. For entrapped air bubbles, adequate defoamers must be used.
Lack of hiding power	Hiding power is a function of the pigment concentration. Insufficient hiding pigment, too thin coating and measurement of hiding when paint is wet are some of the causes. It is noticed that the wet paint appears to hide adequately but on drying of the paint, the film starts showing lack of hiding which is due to the fact that the refractive index of film changes as it dries.	Optimize pigment content and film thickness. Consider adjusting refractive index of binder.
Cissing, retrogation and fat edges	All these defects are related to surface tension. When the surface is not adequately wetted, defects in various sizes result. When the surface on	Use better wetting agents, reduction in surface tension by combination of solvents, better surface preparation by more effective degreasing agents, removal of oil

	which wetting doesn't take place is small (1-2 mm diameter) we call it cissing. When its 3-10 mm diameter we call it cratering. When the exposed surface is irregularly shaped, we call it retrogation. In this case a fat front moving away from the painted surface leaving behind a very inadequately coated surface or even completely exposed bare surface can be seen. All these defects is due to improper surface preparation, presence of oil and grease on the surface or presence of oil droplets in the compressed air. Fat edges result when the entire surface is rejecting the coating and the coating is retrograding away from it. This leads to fat edges formation meaning accumulation of excessive coating material on the edges. When the	droplets from compressed air, slight increase in viscosity of the paint and use of faster evaporating solvents.

	fat edge becomes too thick, it is also called as picture framing.	
Blistering, bubbling	Blistering is due to too rapid rise in temperature of the surface. If the solvent doesn't evaporate quicker than the surface drying it forms a bubble.	Increase flash off time. If cure rate is very rapid, the concentration and type of curing catalyst have to be reviewed.
Bleeding of pigments in case of coloured coatings.	Bleeding is caused by solubility of pigments in resins and solvents.	Choose non-bleeding pigments. Optimize solvent system.
Clouding	Clouding is the appearance of a milky, cloudy, or whitish veil on the coating surface, reducing gloss and transparency. It is due to the moisture condensation, use of highly volatile or incorrect solvent blend, high humidity, low substrate temperature, too thick coat	Slow down solvent evaporation Improve environmental conditions including humidity and temperature Apply thinner coats and ensure proper atomization in spray applications.
Rainbow effect in transparent films	This is due to micro-precipitation of one of the components of resin due to incompatibility. The	Use compatible resins.

	difference in refractive index of the two phases leads to optical interference that results in multiple color development. This defect is most likely caused by using a combination of resins, that are not soluble in each other but exists as a solution due to presence of a common solvent. Once the solvent departs, the two resins separate from one another causing rainbow effect.	
Porosity	This is common in distempers, plastic emulsion paints, intermediate coats and primers if all are formulated at high PVC. If the PVC exceeds CPVC, porosity results. Another reason is the use of extenders with very fine pore size, size that can be filled by water or solvent but not by the binder or emulsion particles. Thus, when coating is applied, the	Use proper pigment volume concentration and complete avoidance of porous extenders. PVC should be about 60-70% of CPVC so that all the space between pigmentary particles is filled up only by the binder.

	pores of the extenders are filled up with water or solvent when the film is applied and allowed to dry, the water or solvent evaporates leaving behind porosity in the film. Porous film has poor resistance to staining, has high permeability for water, gases, and corrosive ions and is a poor protector of the surface. It also has poor washability.	
Bronzing	This defect is typical of certain pigments that change their crystalline form during use. Often found in deep blue shades prepared with phthalocyanine pigments. This defect may not be seen immediately after application but may develop in about a few months to a year. In this case, the surface acquires a metallic look resembling bronze metal-that has the typical red colour imparted by copper.	Use of better stabilized pigments, use of correct wetting and dispersing agents. Careful selection of solvent so that the pigment doesn't show bleeding tendency.

	In the case of other organic pigments, the crystal phase transformation may be slower and may take more time to appear.	
Chalking and erosion	Chalking means easy removal of pigment from the painted surface just by slight touch. This may occur in freshly painted surface due to inadequate binder content, incomplete curing, and use of dead lime in lime washes. This may also occur in older painted surfaces due to degradation of the binder due to weathering.	Remedies for the fresh paint are incorporation of enough binder, allowing full cure for cementaceous coatings, use of properly calcined unexposed lime in lime finishes, etc. For long range improvement, use better binders, UV stabilization by additives. Use abrasion resistant fillers like calcined alumina, quartz, silicon carbide.
Peeling/loss of adhesion	This may occur due to improper surface preparation, variable coat thickness, and uneven exposure to sunlight, or may be due to total mismatch between the surface and the coating or failure of inter-coat adhesion.	Complete removal of loose rust or any other deposits prior to coating by sanding and cleaning. Surface must be free from dirt and oil. Systems with very high polymer Tg can

		potentiate the problem. Careful examination of the location of failure across the film (between substrate and primer, primer and topcoat) may reveal the exact nature of the problem. Flexibilizing agents and adhesion agents (polysiloxanes and phosphates) can be used in many cases mainly in PU-based systems.
Flaking	This is most common with architectural furnishes, paints for ceiling, oil bound distempers etc. most of these products are formulated with high PVC. When the binder content is too low the film loses integrity and flakes out.	Provide adequate binder content in the formulation. If one chooses to use low binder, techniques such as blending of a range of particle size of pigments extenders to provide highest degree of packing may be useful.
Fading/ loss of color	Light fastness of pigments must be checked before use.	Use pigments that are light fast. Addition of UV and light stabilizers can

		help to improve the light fastness of cheaper pigments.
Water spots	Freshly painted, not yet cured coating Layer thickness too high, drying time too short Curing error or hardener no longer usable Use of unsuitable thinner	Observed processing instructions according to product information.
Grinding grooves	Sanding with too coarse sandpaper Soft, elastic substrates Top coat applied too thinly	Solvent test Use specified abrasive paper Apply normal layer thickness
Lap marks	They are visible lines or differences in texture and sheen that occur when paint is applied unevenly or dries inconsistently, especially on large surfaces or when work is interrupted.	Work using the "wet-on-wet" method, applying paint continuously until the surface is covered. Use rollers or brushes that ensure even application Avoid applying paint in high-temperature or high-humidity conditions that speed up drying.

20

Bibliography

This chapter mentions all the references the data is taken from.

Bibliography

1. S. Patil, "Testing of Paints", Current Awareness Service, 1993
2. A. Bragatto, A. Costa, B. Da Silva, B. Dario, F. Rosa, H. Da Silva, J. Barne, J. Santos, L. De Almeida, M. Passarelli, N. De Paula, N. Dantas, R. Da Silva Salvato, R. Da Silva, R. Pagani, R. Ferreira, T. Magri, "Easy guide to paint technology – A quick and practical handbook to formulating paints and emulsions", 3rd edition, Editora Edgard Blucher Ltda., 2024
3. D. C. Webster, "A guide to coatings calculations"
4. V. C. Malshe, M. Sikchi, "Basics of Paint Technology", 2002
5. C. Schoff, "Surface defects: diagnosis and cure", 1998
6. C. Schoff, "Craters and other coatings defects: Mechanisms and analysis", Springer International Publishing AG, 2017
7. M. Ionescu, "Chemistry and Technology of Polyols for Polyurethane", Rapra Technology Ltd., 2005
8. R. W. Hein, "Driers for waterborne coatings", Technology Forum Article, Vol. 70 (886), 1998
9. S. Erich, O. Koc, M. Michel, C. Thomas, L. Ven, H. Huinink, J. Flapper, F. Duivenvoorde, O. Adan, "The influence of calcium and zirconium based secondary driers on drying solvent borne alkyd coatings", Polymer, 2017
10. T. Mezger, "Applied rheology", Anton Paar GmbH, 2014.
11. R. K. Bansal, "A textbook of fluid mechanics and hydraulic machines", Laxmi Publications Pvt. Ltd., 9th edition.
12. B. Abderezzak, "Introduction to Transfer phenomena in PEM fuel cell", Elsevier Ltd., 2018.
13. A. R. Blythe, "Electrical resistivity measurements of polymer materials", Polymer Testing, Elsevier Applied Science Publishers Ltd., 4 (1984), 195-209
14. J. Sun, H. Gokturk, D. Kalyono, "Volume and surface resistivity of low density polyethylene filled with stainless steel fibers", Journal of Materials Science, Chapman and Hall, 28 (1993), 364-366.

15. I. Sahalianov, S. Singh, K. Tybrandt, M. Berggren, I. Zozoulenko, "The intricnsic volumetric capacitance of conducting polymers: pseudo-capacitor or double-layer supercapacitors?", RSC Advances, 9 (2019), 42498-42508
16. G. Wypych, "Handbook of fillers", 4th edition, ChemTec Publishing, 2016
17. M. Mulder, "Basic principles of membrane technology", 2nd edition, Kluwer Academic Publishers, 1996
18. M. Beer, P. Wood, J. Weisz, "A simple and rapid method for evaluation of Mark-Houwink-Sakurada constants of linear random coil polysaccharides using molecular weight and intrinsic viscosity determined by high performance size exclusion chromatography: application to guar galactomannan", Carbohydrate Polymers, Elsevier, 39 (1999), 377-380
19. J. Rauschkolb, B. Ribeiro, T. Feiden, B. Fischer, T. Weschenfelder, R. Cansian, A. Junges, "Parameter estimation of Mark-Houwink Equation of polyethylene glycol (PEG) using molecular mass and intrinsic viscosity in water", Biointerface Research in Applied Chemistry, 12, 2 (2022), 1778-1790
20. V. Halabalova, L. Simek, J. Dostal, M. Bohdanecky, "Note on the relation between the parameters of the Mark-Houwink-Kuhn-Sakurada Equation", International Journal of Polymer Analysis and Characterization, 9 (1), 65-75
21. V. Shah, "Handbook of plastics testing and failure analysis", 3rd edition, John Wiley and sons, Inc, 2007
22. P. Ortega, R. Anyszka, Y. Morishita, R. Ronza, A. Blume, "Determination of the crosslink density of silica-filled styrene butadiene rubber compounds by different analytical methods", Polymer Bulletin, 2023
23. M. Pieruccini, M. Fernandez, G. Vozzolo, M. Ximenis, R. Aguirresarobe, J. Vega, "Dynamic heterogeneity and cooperativity in polyurethane-based vitrimers", Polymer, 324 (2025) 128192

24. V. Villani, V. Lavallata, "The theories of rubber elasticity and the goodness of their constitutive stress-strain equations", Physchem, MDPI, 2024
25. R. Arjanggi, J. Kansedo, "Recent advancement and prospective of waste plastics as biodiesel additives: A review", Journal of the Energy Institute, 2019.
26. Fink "Reactive polymers fundamentals and applications", Elsevier Inc., 2013
27. B. Roughton, J. White, K. Camarda, R. Gani, "Simultaneous design of ionic liquids and azeotropic separation processes", 21st European Symposium on Computer Aided Process Engineering-ESCAPE 21, 2011
28. "Adhesive Technology Handbook", Elsevier Inc., 2015
29. F. Smedes, R. Geertsma, T. Zande, K. Booij, "Polymer-water partition coefficients of hydrophobic compounds for passive sampling: Application of cosolvent models for validation", Environmental science and technology, 43 (2009), 18, 7047-7054
30. E. Tehrany, S. Desobry, "Partition coefficient of migrants in food simulants/ polymers system", 101 (2007), 4, 1714-1718
31. B. Li, Z. Wang, Y. Bai, "Determination of the partition and diffusion coefficients of five chemical additives from polyethylene terephathalate material in contact with food simulants", 21 (2019), 100332
32. N. Silva, L. Ferreira, P. Madeira, J. Teixeira, V. Uversky, B. Zaslavsky, "Linear relationships between partition coefficients of different organic compounds and proteins in aqueous two-phase systems of various polymer and ionic compositions", Polymers, 12 (2020), 7, 1452.
33. Q. Wang, L. Xing, "Determination of fracture toughness K_{IC} by using the flattened Brazilian disk specimen for rocks", Engineering fracture mechanics, 64 (1999), 2, 193-201
34. E. Tarani, I. Arvanitidis, D. Christofilos, D. Bikiaris, K. Chrissafis, G.. Vourlias, "Calculation of the degree of crystallinity of HDPE/GNPs nanocomposites by using various

experimental techniques: a comparative study", Journal of Material Science, 58 (2023), 1621-1639

35. A Faye, M. Leduc, J. Brisson, "Crystallization control of etherethersulfone copolymers by regular insertion of an allyl functionality", Polymer Chemistry, 5 (2014), 2548
36. Y. Furushima, M. Nakada, M. Murakami, T. Yamane, A. Toda, C. Schick, "Method for calculation of the lamellar thickness distribution of not-reorganized linear polyethylene using fast scanning calorimetry in heating", Macromolecules
37. J. Weeks, "Melting temperature and change of lamellar thickness with time for bulk polyethylene", Journal of Research of the National Institute of Standards and Technology, 1 (1963), 67A(5), 441-451
38. S. yang, J. Lei, G. Zhong, J. Xu, Z. Li, "Role of lamellar thickening in thick lamellae formation in isotactic polypropylene when crystallizing under flow and pressure", Polymer, 179 (2019), 28, 121641
39. M. Mulder, "Basic Principles of Membrane Technology", second edition, Kluwer Academic Publishers, 1996
40. J. Fink, "Petroleum Engineer's Guide to Oil Field Chemicals and Fluids", Elsevier, 2012
41. B. Morris, "The Science and Technology of Flexible Packaging", Elsevier, 2017
42. S. Dey, "Enhancement in hydrophobicity of low rank coal by surfactants- A critical overview", Fuel Processing Technology, 94 (2012), 151-158
43. D. Swinehart, "The Beer-Lambert Law", Journal of Chemical Education, 39 (1962), 7, 333
44. T. Mayerhofer, S. Pahlow, J. Popp, "The Bouguer-Beer-Lambert Law: Shining light on the Obscure", Chemphyschem, 21 (2020), 18, 2029-2046
45. J. Vlachopoulos, "Die swell and normal stresses: An explaination", Rubber Chemistry and Technology, 51 (1978), 1, 133-138

46. D. Karthik, J. Militky, Y. Wang, M. Venkataraman, "Joule heating of carbon-based materials by carbonisation of para-aramid fibrics", Journal of Carbon Research, 9 (2023), 1, 23
47. R. Wellen, E. Canedo, "On the Kissinger equation and the estimate of activation energies for non-isothermal cold crystallization of PET", Polymer Testing, 40 (2014), 33-38
48. S. Vyazovkin, "Kissinger method in kinetics of materials: Things to beware and be aware of", Molecules, 25 (2020), 2813
49. Y. Cho, M. Shim, S. Kim, "Thermal degradation kinetics of PE by the Kissinger equation", Materials Chemistry and Physics, 52 (1998), 94-97
50. Y. Shangguan, F. Chen, E. Jia, Y. Lin, J. Hu, Q. Zheng, "New insight into time-temperature correlation for polymer relaxations ranging from secondary relaxation to terminal flow: Application of a universal and developed WLF equation", 9 (2017), 11, 567
51. O. Starkova, A. Gagani, C. Karl, I. Rocha, J. Burlakovs, A. Krauklis, "Modelling of environmental ageing of polymers and polymer composites- durability prediction methods", Polymers, 14 (2022), 907
52. J. Zhu, Y. Deng, P. Chen, G. Wang, H. Min, W. Fang, "Prediction of long-term tensile properties of glass fiber reinforced composites under acid-base and salt environments", Polymers, 14 (2022), 3031
53. C. Avenel, O. Raccurt, J. Gardette, S. Therias, "Accelerated aging test modelling applied to solar mirrors", Materials Degradation, 3 (2019), 27
54. A. Andrady, J. Pegram, Y. Tropsha, "Changes in carbonyl index and average molecular weight on embrittlement of enhanced-photodegradable polyethylenes", Journal of Environmental Polymer Degradation, 1 (1993), 3, 171-179
55. E. Thyni, M. Walfridson, "Automation of carbonyl index calculations for fast evaluation of microplastics degradation", Degree Project in Technology, KTH Royal Institute of Technology, 2022.

56. J. Almond, P. Sugumaar, M. Wenzel, G. Hill, C. Wallis, "Determination of the carbonyl index of polyethylene and polypropylene using specified area under band methodology with ATR-FTIR spectroscopy", e-Polymers, De Gruyter, 20 (2020), 369-381.
57. S. Goh, "Thermoanalytical studies of rubber oxidation: correlation of activation energy, isothermal induction time and oxidation peak temperature", Thermochimica Acta, 77 (1984), 1-3, 275-280
58. J. Zmpitas, J. Gross, "Modified Stokes-Einstein Equation for molecular self-diffusion based on entropy scaling", Industrial and Engineering Chemistry Research, 60 (2021), 4453-4459
59. S. Khrapak, A. Khrapak, "Stokes-Einstein relation in different models of water", molecules, 29 (2023), 23, 5587
60. M. Tibbitt, "Rubber elasticity", Networks and Gels, 2022
61. G. Zaikov, A Haghi, E Klodzinska, "Materials Science and Engineering, Volume II Physiochemical concepts, Properties, and Treatments", 1st edition, Taylor and Francis Group, 2014
62. G. Biresaw, K. Mittal, "Surfactants in Tribology, Volume 4", 1st edition, Taylor and Francis Group, 2015
63. M. Kamran, "Fuel cell, Renewable Energy Conversion Systems", Academic Press, 2021, 221-242
64. R. Haj-Kacem, N. Ouerfelli, J. Herrae, M. Guettari, H. Hamda, M. Dallel, "Contributions to modelling the viscosity Arrhenius-type equation for some solvents by statistical correlations analysis", Fluid Phase Equilibria, 383 (2014), 11-20
65. E. Ike, S. Ezike, "Estimation of viscosity Arrhenius pre-exponential factor and activation energy of some organic liquids", International Journal of Recent Research in Physics and Chemical Sciences, 5, 1 (2018), 18-26
66. Z. Tao, Z. Youcai, E. Nyankson, "Composite materials production using waste plastic taken from municipal solid waste and its landfill", (2023), 187-213.
67. https://uk.ppgrefinish.com/en/paint-defects/blushing/# Accessed on 06/09/2025

68. https://www.zuelch.de/en/services/painting-defects/ Accessed on 09/09/2025
69. https://www.rheologylab.com/articles/coatings-paints-inks/coating-defects/ Accessed on 09/09/2025
70. https://www.teknos.com/industrial-coatings/blog/7-common-industrial-coating-defects Accessed on 09/09/2025
71. https://coatinginstitute.org/en/paint-defects/ Accessed on 09/09/2025
72. https://www.britishpaints.in/index.php/paint-defects Accessed on 09/09/2025
73. https://www.crownpaints.co.ke/paint-defects/ Accessed on 09/09/2025
74. http://www.smartchem.com.ar/pdf/MCR-Drier-Dosage.pdf Accessed on 16/09/2025
75. http://www.specialchem.com/plastics/guide/dielectric-constant Accessed on 24/09/2025
76. https://www.specialchem.com/plastics/guide/dissipation-factor Accessed on 24/09/2025
77. https://www.study.com/skill/learn/how-to-calculate-the-capacitance-of-a-parallel-plate-capacitor-of-given-its-geometry-explanation.html. Accessed on 25/09/2025
78. https://www.ncbi.nlm.nih.gov/sites/books/NBK441935/ Accessed on 09/11/2025
79. https://ultraaqua.com/blog/uv-dose/ Accessed on 09/11/2025
80. https://www.alpha-cure.com/news-article/the-right-dose-for-uv-disinfection Accessed on 09/11/2025
81. https://www.versaperm.com/Ficks%20Law.php Accessed on 09/11/2025
82. https://en.wikipedia.org/wiki/Moisture_vapor_transmission_rate Accessed on 09/11/2025
83. https://www.chem.uci.edu/~dmitryf/manuals/Fundamentals/SLS Technologies GPC-SEC Explained.pdf Accessed on 09/11/2025
84. https://www.malvernpanalytical.com/en/learn/knowledge-center/insights/what-is-the-mark-houwink-equation-and-how-

do-you-use-it-for-polymer-characterization Accessed on 09/11/2025
85. https://wiki.anton-paar.com/en/intrinsic-viscosity-determination/ Accessed on 09/11/2025
86. https://satheejee.iitk.ac.in/article/physics/physics-relation-between-youngs-modulus-and-bulk-modulus/ Accessed on 09/11/2025
87. https://iupac.org/wp-content/uploads/2017/12/IUPAC_PolymEdu_Shortcourse_9ppt_AikHweeEng.pdf Accessed on 09/11/2025
88. https://en.wikipedia.org/wiki/Fracture_toughness Accessed on 09/11/2025
89. https://www.nde-ed.org/Physics/Materials/Mechanical/NotchToughness.xhtml Accessed on 09/11/2025
90. https://en.wikipedia.org/wiki/Heat_capacity Accessed on 09/11/2025
91. https://en.wikipedia.org/wiki/Thermal_conductivity_and_resistivity Accessed on 09/11/2025
92. https://en.wikipedia.org/wiki/Thermal_diffusivity Accessed on 09/11/2025
93. https://physics.stackexchange.com/questions/675134/what-is-the-difference-between-thermal-conductivity-and-thermal-diffusivity Accessed on 09/11/2025
94. https://en.wikipedia.org/wiki/Crystallinity Accessed on 09/11/2025
95. https://serc.carleton.edu/research_education/geochemsheets/BraggsLaw.html Accessed on 09/11/2025
96. https://www.vedantu.com/physics/braggs-law Accessed on 09/11/2025
97. https://wiki.anton-paar.com/en/x-ray-diffraction-xrd/ Accessed on 09/11/2025
98. https://testbook.com/physics/braggs-law Accessed on 09/11/2025

99. https://en.wikipedia.org/wiki/Fick%27s_laws_of_diffusion Accessed on 09/11/2025
100. https://www.nanoscience.com/techniques/tensiometry/contact-angle-measurements-and-wettability/ Accessed on 09/11/2025
101. https://www.biolinscientific.com/blog/what-is-spreading-coefficient Accessed on 09/11/2025
102. https://ethz.ch/content/dam/ethz/special-interest/chab/icb/shih-lab-dam/documents/IEM-lecture/Lecture_2024/Notes_2024/Lecture_2_2024.pdf Accessed on 09/11/2025
103. https://en.wikipedia.org/wiki/Capillary_number Accessed on 09/11/2025
104. https://en.wikipedia.org/wiki/Washburn%27s_equation Accessed on 09/11/2025
105. https://spie.org/samples/PM229.pdf Accessed on 09/11/2025
106. https://www.specialchem.com/coatings/guide/opacity Accessed on 09/11/2025
107. https://www.duluxprotectivecoatings.com.au/media/1581/531_handy_calculations-vol_solids_spread-rate.pdf Accessed on 09/11/2025
108. https://ansyshelp.ansys.com/public/account/secured?returnurl=/Views/Secured/corp/v251/en/poly_pm/pm_ug_sec_guidelines_weissenberg.html Accessed on 09/11/2025
109. https://www.kruss-scientific.com/en/know-how/glossary/method-according-to-zisman Accessed on 09/11/2025
110. https://www.tainstruments.com/pdf/literature/RN11.pdf Accessed on 09/11/2025
111. https://www.tainstruments.com/pdf/literature/TA144.pdf Accessed on 09/11/2025
112. https://goldbook.iupac.org/terms/view/12821 Accessed on 09/11/2025
113. https://en.wikipedia.org/wiki/Williams%E2%80%93Landel%E2%80%93Ferry_equation Accessed on 09/11/2025

114. https://www.mt.com/int/ar/home/applications/Application_Browse_Laboratory_Analytics/Application_Browse_thermal_analysis/oxidation-induction-time.html Accessed on 09/11/2025
115. https://support.nanotempertech.com/hc/en-us/articles/23851436308625-The-Stokes-Einstein-Equation Accessed on 09/11/2025
116. https://www.fidabio.com/blogs/stokes-radius-and-stokes-einstein-equation Accessed on 09/11/2025
117. https://en.wikipedia.org/wiki/Poisson's_ratio Accessed on 05/01/2026

www.ingramcontent.com/pod-product-compliance
Lightning Source LLC
LaVergne TN
LVHW050934210926
848482LV00013B/613